Electronic Circuits

Volume 1.3

Disclaimer

Electronic Circuits

Volume 1.3

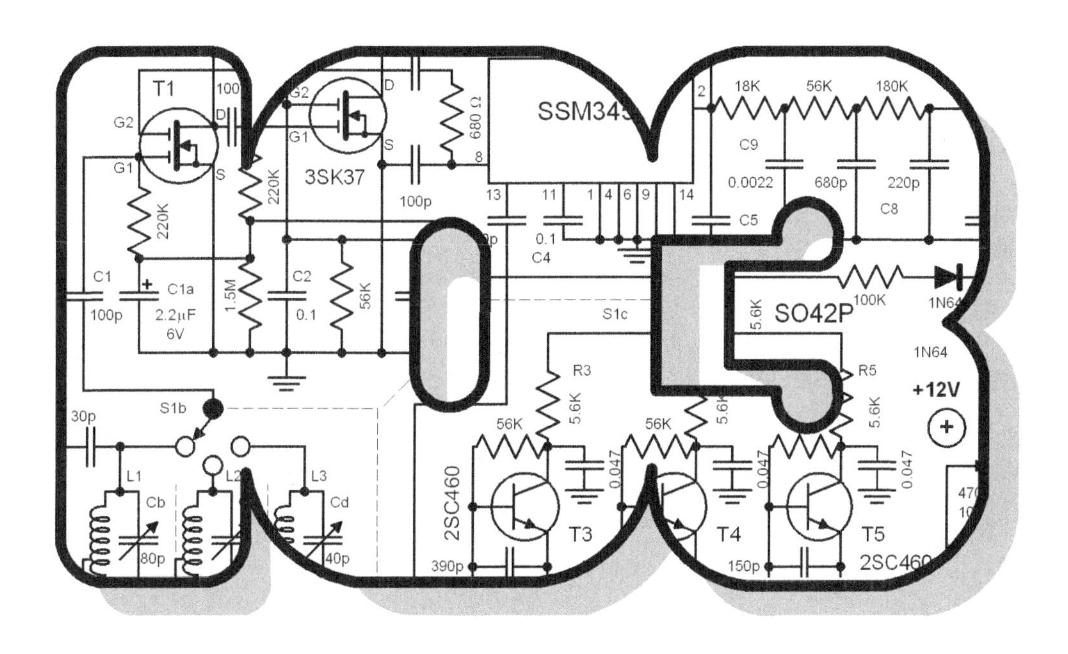

Electronic Circuits Volume 1.3

Published by the

www.intellin.org

First year of publication 2008
Published in the USA

Disclaimer:

The electronic circuits, software or related documentation in this book are NOT designed nor intended for use (whether free or sold) as on-line control equipment in hazardous environments requiring fail-safe performance, such as, but not limited to, in the operation of nuclear facilities, aircraft navigation or communication systems, air traffic control, direct life support machines or weapons systems in which the failure of the hardware or software could lead directly to death, personal injury, or severe physical or environmental damage ("high risk activities")

The author(s) and publisher(s) take no responsibility for damages or injuries of any kind that may arise from the use or misuse of the electronic circuits in this collection.

The author(s) and publisher(s) specifically disclaim any express or implied warranty or fitness for high risk activities. The electronic circuits, software and related documentation are without warranty of any kind. The author(s) and publisher(s) expressly disclaim all other warranties, express or implied, including, but not limited to, the implied warranties of merchantability and fitness for a particular purpose. Under no circumstances shall the author(s) and publisher(s) be liable for any incidental, special or consequential damages that result from the use or inability to use the circuits and software or related documentation, even if he has been advised of the possibility of such damages.

ISBN 1-4196-9005-1

ISBN13 (EAN13) 9781419690051

Congratulations for having the fourth volume of ready-to-apply electronic circuits. With this book, you got the luxury of being able to design and assemble electronic modules fast and worry free. It is a sure way to optimize satisfaction in your hobby. If you are a professional electronic designer, it will help you beat the competition. Speed, efficiency, short development periods, error-free, user and maintenance friendly: these are the factors critical for success. This invaluable book filled with 103 practical ideas will help you beat project deadlines. Make your ideas work!

Make your creativity pay! All that JUST IN TIME!

informative...

practical...

professional...

versatile...

Acknowledgments

Many Thanks to...

E. Mischa (Optical Recognition)
D. Salinger (Electronics)
Peter Schmidt (Cybernetics)
N. Lay (Robotics)

INTRODUCTION

This collection contains 103 practical electronic circuits that are grouped in nine general applications. Since most of the electronic circuits are not limited to a single application, a circuit may have found its way into another group. This is one proof of the versatility of the circuits. Creativity needs versatility . You can combine several circuits into one large module to create a powerful electronic device specially designed for your exclusive project.

The table of contents lists the groups and titles of the circuits. The number of circuits available in the application group is next to the right of the application group's name. The page numbers are shown next to the right of the circuit list.

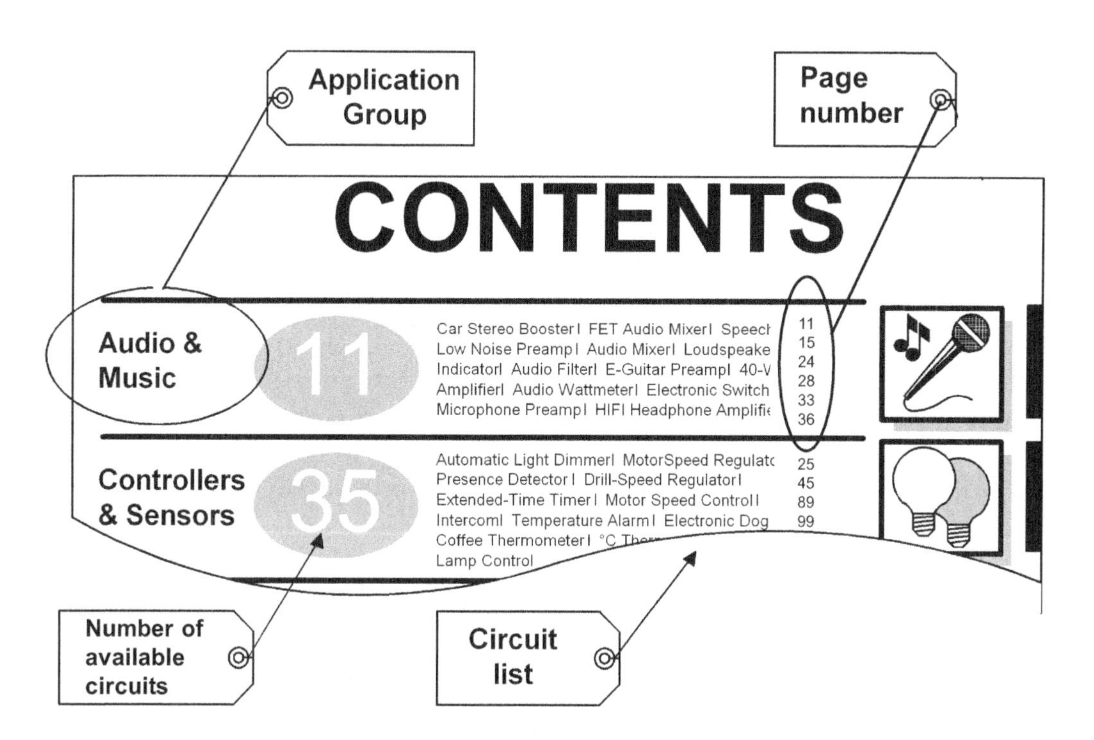

The transistors used in the circuits have more than one possible replacements. The pin designations are also shown in details. This feature can help avoid unnecessary delays. The pins shown are either in the bottom view or front view of the transistor unless otherwise noted. Large transistors which cannot or not planned to be installed directly on the PCB must be installed on a heatsink. A dashed circle around a transistor means that the transistor must be heatsinked.

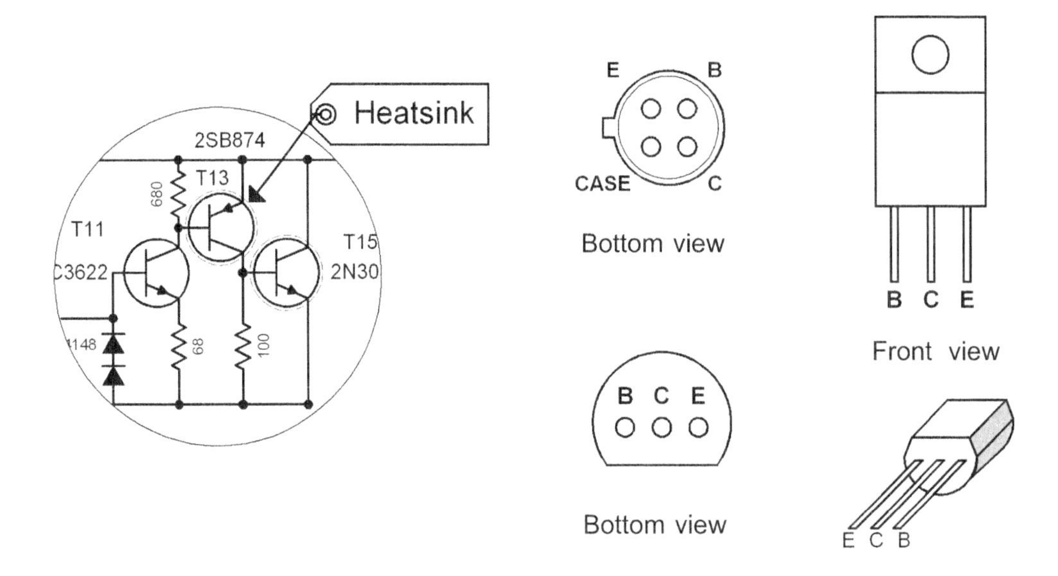

The capacitor values are given in microfarad unless otherwise specified. Electrolytic or polarized capacitors are marked with a plus sign in the diagram. This plus sign coincides with the capacitor's positive polarity in the circuit. Additionally, their voltage ratings are also given. Nonpolar capacitors are ceramic types and rated with 50 volts.

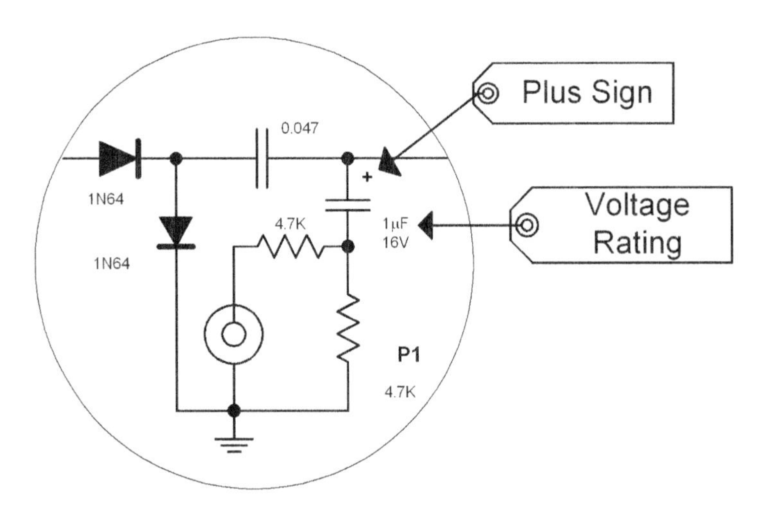

The resistor values are given in ohms (Ω), rated 1/4 watts and are of carbon film type unless otherwise specified.

CONTENTS

Audio & Music

Radio Frequency

Hobby & Shop

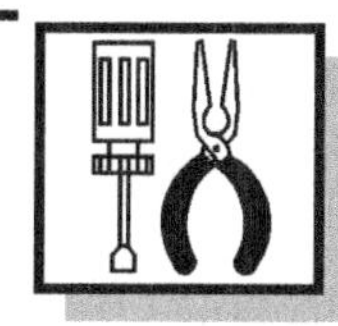

Controllers & Sensors

1 3-CHANNEL AUDIO MIXER

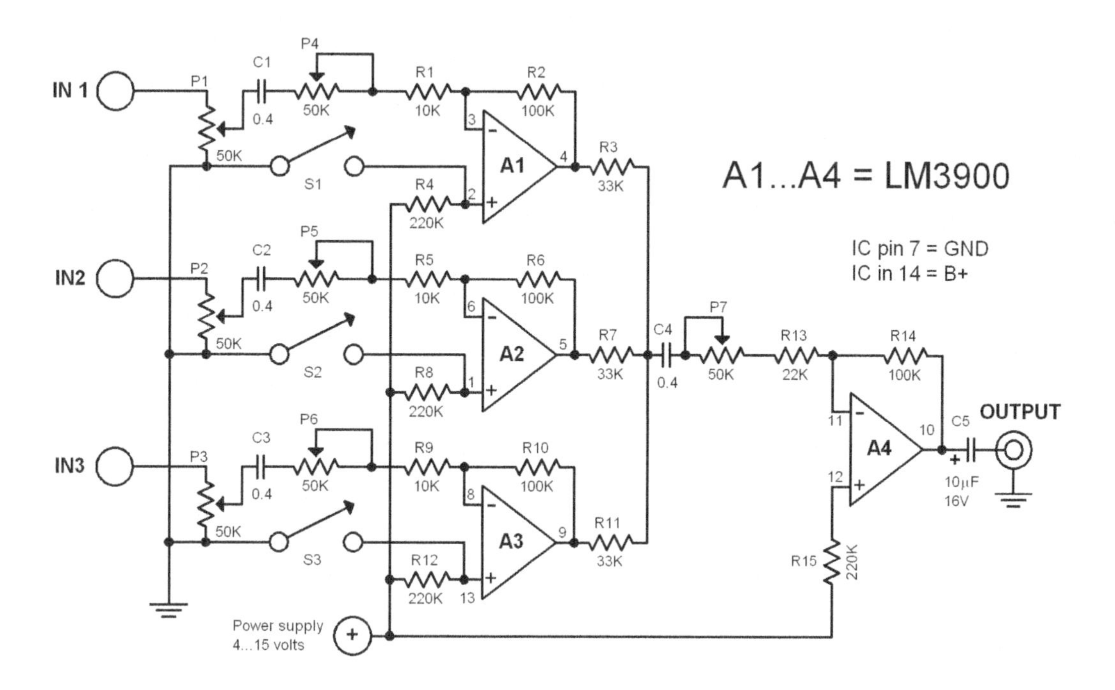

Diagram 1.0 3-Channel Audio Mixer

This audio mixer circuit uses an LM3900 IC. The IC houses four integrated Norton amplifiers. The advantage of using these four opamps is that they only need a single power supply. Since this amplifier circuit is current controlled, the DC bias is dependent on the feedback coupling. The schematic diagram shows inverting AC-Norton amplifiers. The DC output must be set at 50 percent of the power supply. In this case, a maximum output can be achieved without distortion (also called symmetrical limitation through overdrive).

In designing the circuit, you can freely choose the value of the resistor R2 (100K in the above featured circuit). Set the AC voltage amplification factor through the ratio of R2/R1. To set the amplifier gain correctly, choose the value of R4 = 2R2 (double the value of R2).

Diagram 1.0 shows the 3-channel mixer circuit using three Norton-opamps. The input levels can be set by potentiometers P1 to P3. Furthermore, each input level can be trimmed with the help of trimmer pots P4 to P6 to adapt each input to the source. The resistors at the non-inverting inputs of the opamps work as DC bias and set the DC output at 50 percent of the power supply. All three input signals are summed by the fourth opamp A4 through the resistors R3, R7, and R11. The common volume level is controlled through the potentiometer P7.

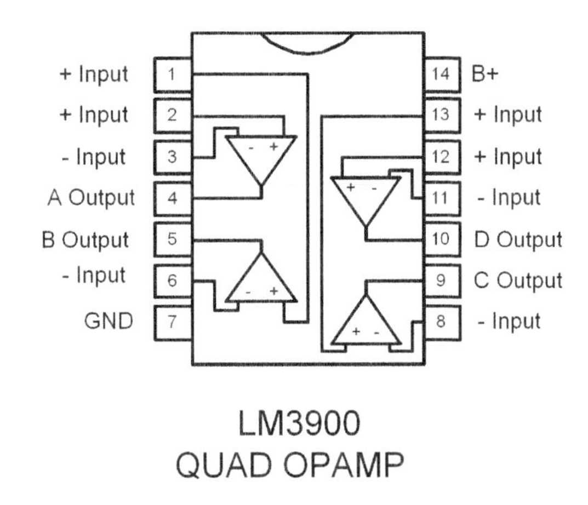

Figure 1.0 Pin Designations

You can switch an input channel on or off through the switches S1 to S3. An input channel is turned off when its switch is closed. It is also possible to replace these mechanical switches with transistor gates. By doing so, you can build an analog multiplexer circuit that can be easily expanded by several inputs.

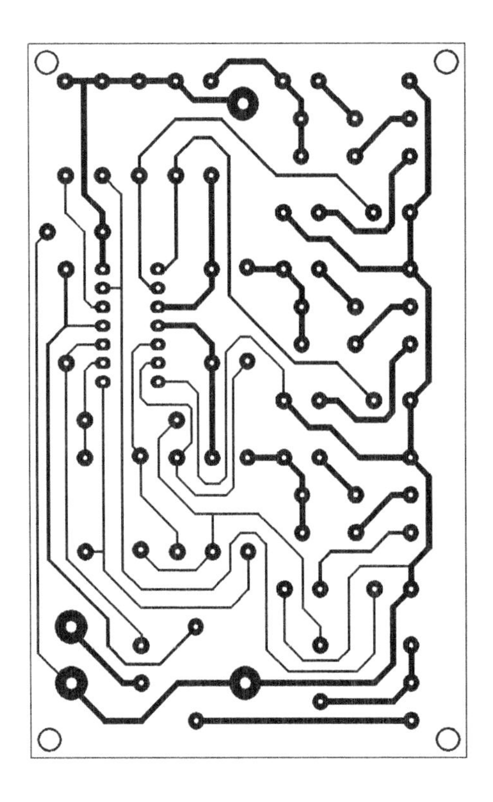

Figure 1.1 Printed Circuit Layout

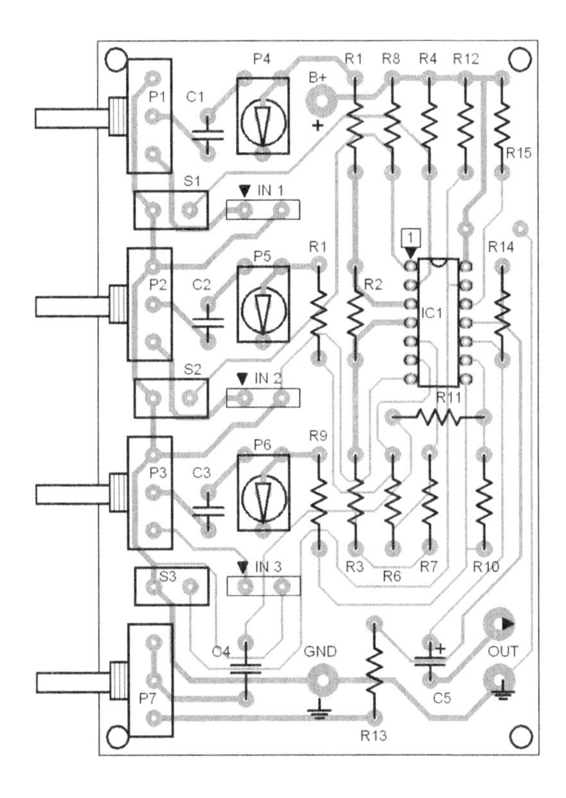

Figure 1.2 Parts Placement Layout

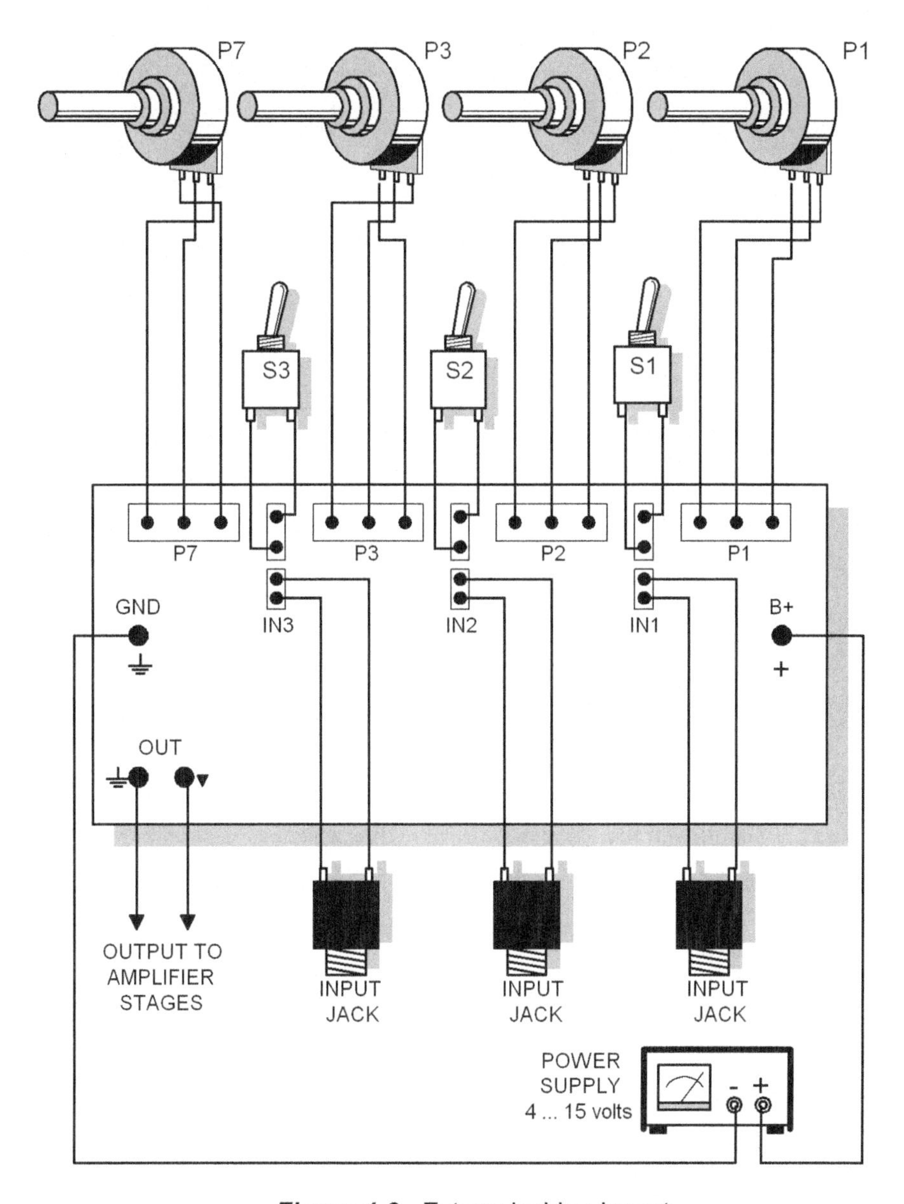

Figure 1.3 External wiring layout

Parts List:	
Resistors:	Capacitors:
R1,R5,R9 = 10K	C1,C2,C3,C4 = 0.4 µF
R2,R6,R10,R14 = 100K	C5 = 10 µF/16volts
R4,R8,R12,R15 = 220K	
R3,R7,R11 = 33K	Potentiometers:
R13 = 22K	P1,P2,P3,P4,P5,P6,P7 = 50K

All resistors are ¼ watts unless otherwise specified.

2 CARDIOPHONE

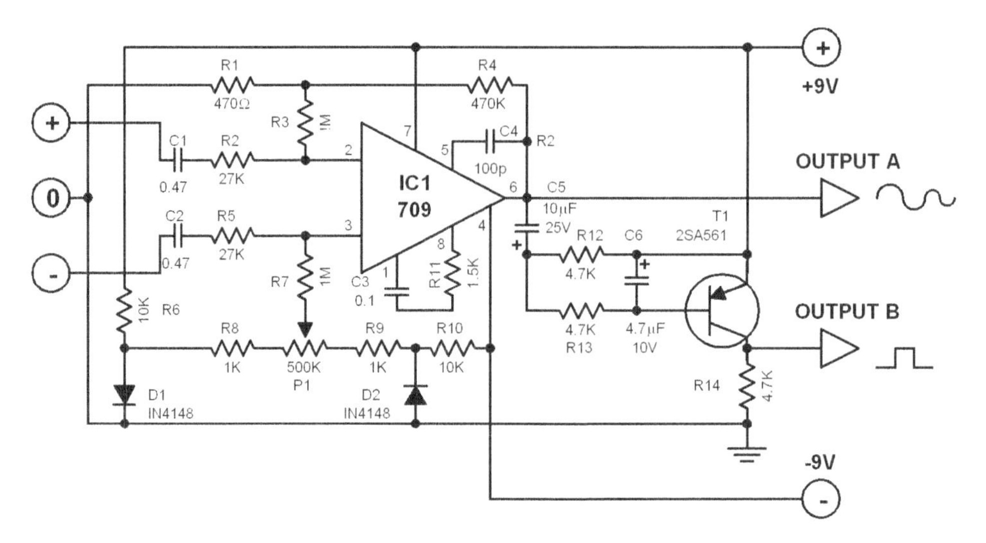

Diagram 2.0 Cardiophone

The human heartbeat can be made audible by using this cardiophone circuit! Diagram 2.0 shows the schematic of such a circuit. It is basically an audio circuit coupled to a probe made specially for the purpose of picking up the electric signal from the human heart.

To get the best signal, place the probe's electrodes to a point close to the heart. The preferred point is just below the left breast with the negative electrode pointing to the left of the sternum.

After constructing the circuit, the output A must be calibrated to null through the potentiometer P1. This is important for the circuit to function properly. The signal coming from output A can then be connected to either a low-frequency amplifier or an oscilloscope.

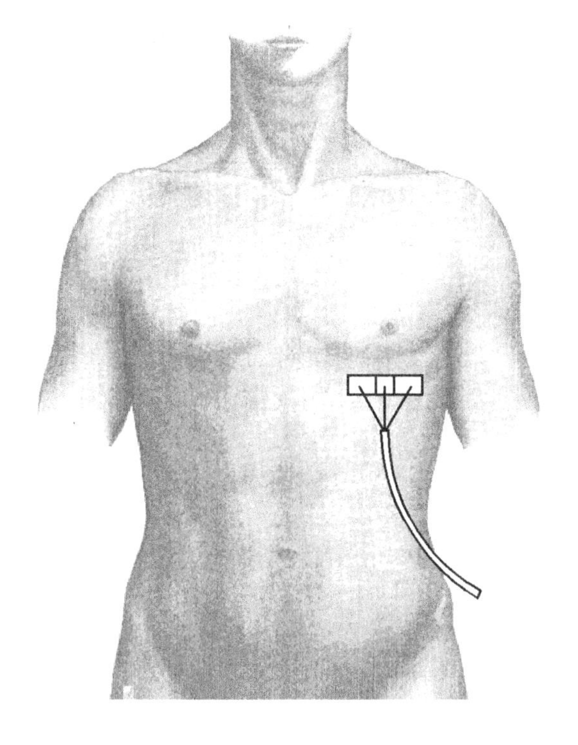

Figure 2.0 Placement of the probe

The signal coming from output B is a square wave in sync with the heart rhythm. This signal can be used to trigger a final amplifier or other circuits. The heartbeat can be heard from the final amplifier's speaker.

The special signal probe is shown on Figure 2.1. The simplest way to make this probe is to use a 1cm x 10 cm blank pcb board. Following the design on Figure 2.1, the non-shaded parts of the pcb board must be etched away. The un-etched copper plate must then be covered with solder to protect it from corrosion and to facilitate good contact with the skin. Take note that two of the probe's electrodes are marked negative and positive respectively. It is of utmost importance to use a shielded twisted pair wire for the cable connecting the probe with the cardiophone circuit.

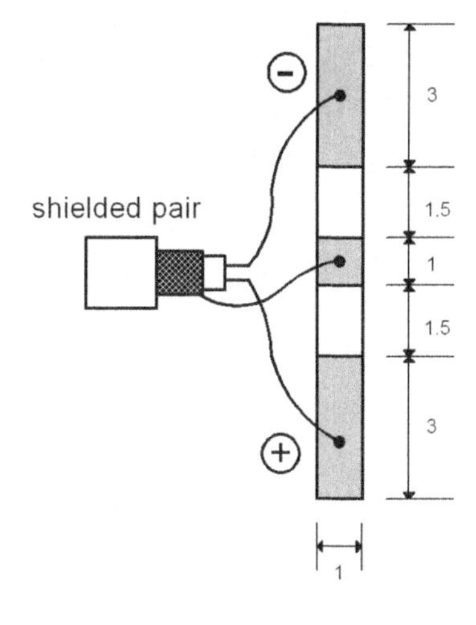

Figure 2.1 Special probe

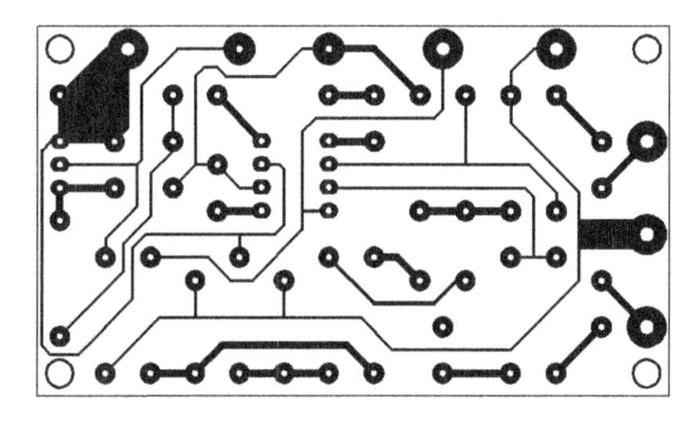

Figure 2.2 Printed Circuit Layout for the Cardiophone

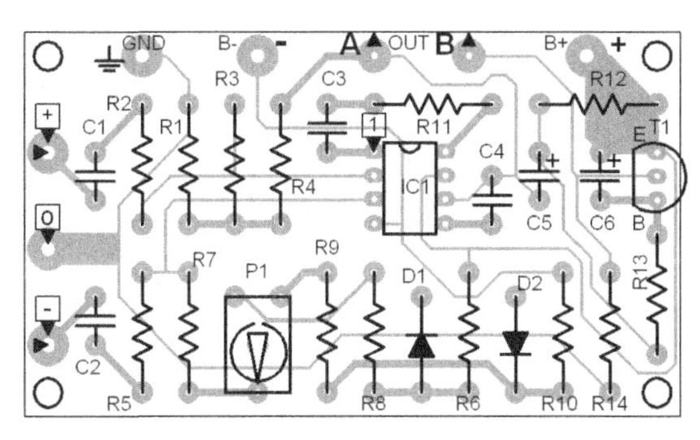

Figure 2.3 Parts Placement Layout for the Cardiophone

Parts List:

Resistors:
R1 = 470 Ω
R2,R5 = 27K
R3,R7 = 1M
R4 = 470K
R6,R10 = 10K
R8,R9 = 1K
R11 = 1.5K
R12,R13,R14 = 4.7K
P1 = 500K trimmer pot
All resistors are ¼ watts unless otherwise specified.

Capacitor:
C1,C2 = 0.47μF/50V ceramic
C3 = 0.1μF
C4 = 10μF/10Volts
C5 = 4.7μF/10Volts

D1,D2 = 1N4148

T1 = 2SA561 Transistor
IC1 = 706 Opamp

3 LM386 AUDIO AMPLIFIER

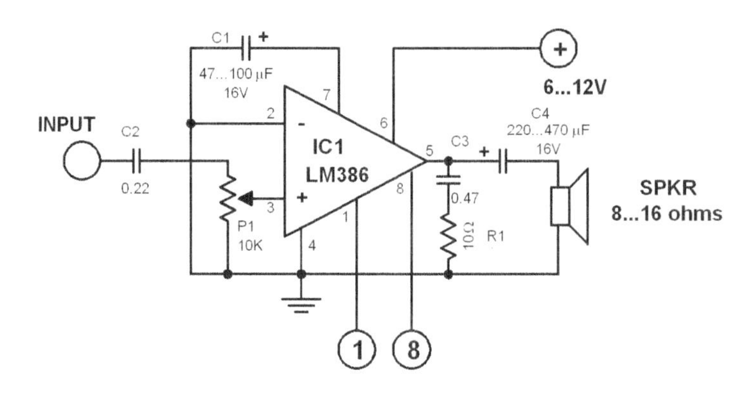

Diagram 3.0 LM386 Audio Amplifier

The integrated chip LM386 is a low power audio frequency amplifier requiring a low level power supply (most often batteries). It comes in an 8-pin mini-DIP package. The IC is designed to deliver a voltage amplification of 20 without external add-on parts. But this voltage gain can be raised up to 200 (Vu = 200) by adding external parts.

The external parts shown on Diagram 3.1 can be selected to get the desired gain. Circuit A will give a voltage amplification of 200. Circuit B will give a gain of around 50. The circuit C is not for voltage amplification but will raise the bass level by about 5 dB. Take note that the circuit C is to be connected between pins 1 and 5 of the IC.

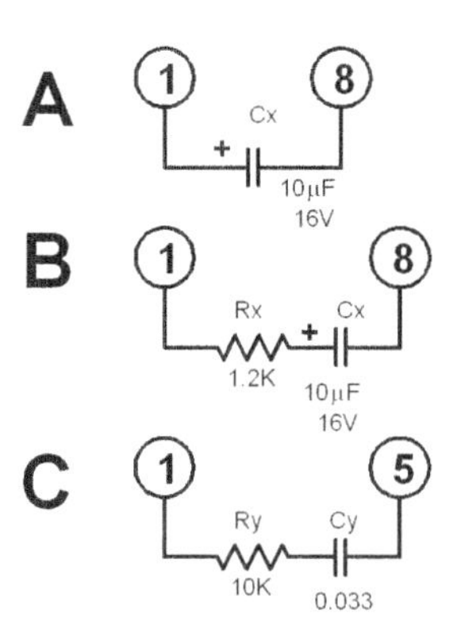

Diagram 3.1 External parts

The power output is around 550 mW at 16 ohm speaker impedance. This audio frequency amplifier is ideal for small battery powered devices.

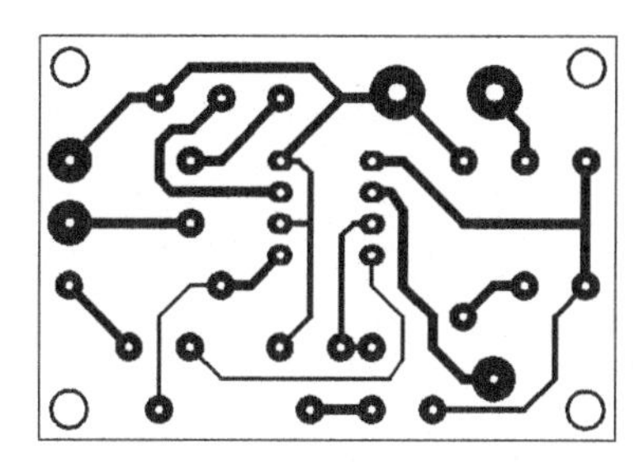

Figure 3.0
Printed Circuit Layout
for the LM386 Audio Amplifier

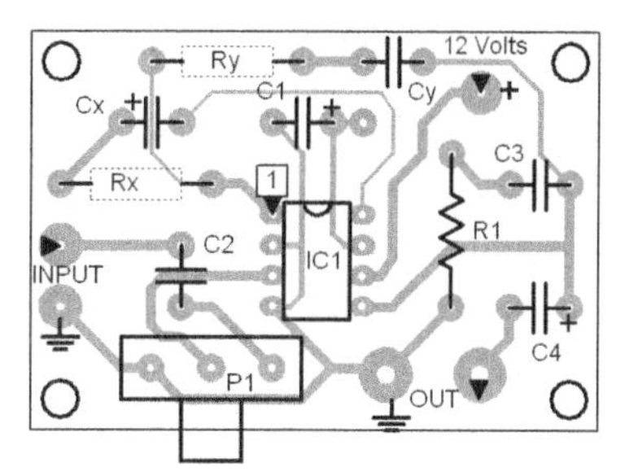

Figure 3.1
Parts Placement Layout
for the LM386 Audio Amplifier

If you use the external circuit A, replace the Rx with a jumper wire in the pcb. If you use the external circuit C, solder the additional resistor and capacitor to the pcb points labeled Ry and Cy, see Figure 3.1.

4 DYNAMIC MIC PREAMP

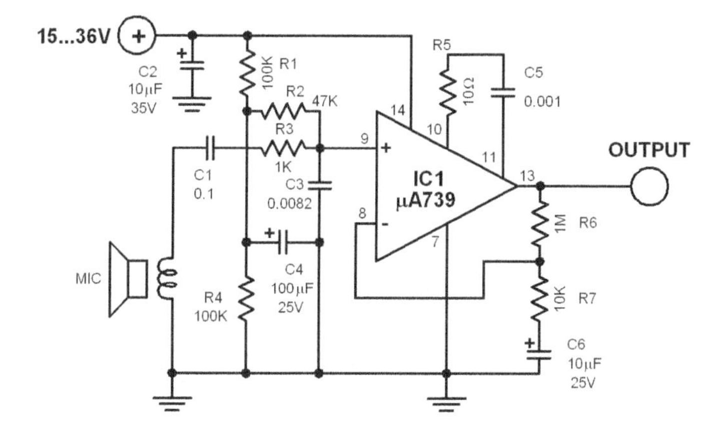

Diagram 4.0 Dynamic Mic Preamp

The above mic preamp uses the low noise IC μA739 IC. The circuit is an example of how a good preamp can be designed for dynamic microphones. The IC houses two identical integrated preamp circuits. The second preamp is used in identical manner for the second channel of the stereo microphone.

Diagram 4.1 shows the pin numbers (in brackets) for the second identical channel. All external parts are identical to those shown in Diagram 2.0

The non-inverting input is biased at about 50% of the power supply. This bias voltage is set by the voltage divider circuit R1 and R4. The point between R1 and R4 is used commonly for both channels.

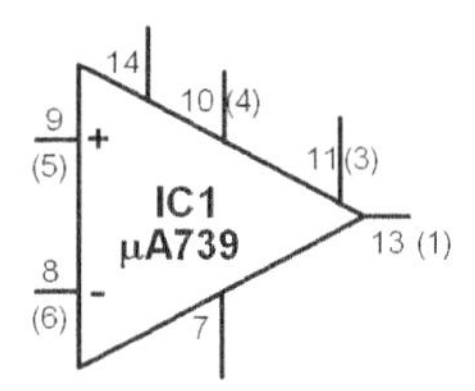

Diagram 4.1 Pin designations for the other preamp inside the IC

The unwanted HF signals coming from the microphone are filtered out by the RC-circuit composed of R3/C4. Frequency compensation is done by the R7/C6 circuit. The values of R7 and C6 were designed to avoid oscillation at the amplification level of 100. The input impedance is about 47K. This means that a normal dynamic microphone gets connected to a high impedance preamp which in turn produces good results. The output impedance is about several hundred ohms.

The maximum peak-peak output voltage is about several volts lower than the supplied power. The frequency range is from 20Hz to 20kHz (-3dB). The upper cutoff frequency is about 80 kHz when the low-pass filter is removed from the circuit. The IC shown can be replaced with TBA231 or SN76131 without changing the external circuit.

Parts List:

Capacitors:
C1 = 0.1 μF
C2 = 10 μF/35volts electrolytic
C3 = 0.0082 μF
C4 = 100 μF/25V
C5 = 0.001 μF
C6 = 10μF/25V electrolytic

Resistors:
R1, R4 = 100K
R2 = 47K
R3 = 1K
R5 = 1Ω
R6 = 1M
R7 = 10K

All resistors are ¼ watts unless otherwise specified.

5 HEADPHONE AMPLIFIER

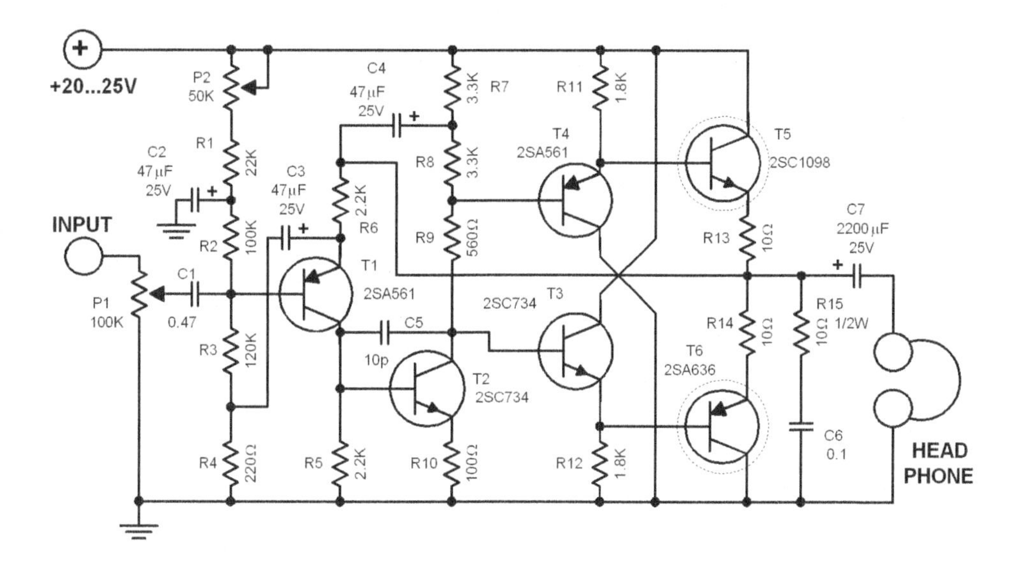

Diagram 5.0 Headphone Amplifier

Typically, a headphone is connected to the loudspeaker output of the final amplifier stages through a voltage divider circuit. However, this simple design has two distinct disadvantages. Firstly, the headphone volume cannot be varied independently from the main speaker when the main speaker is switched on at the same time. Secondly, the voltage divider circuit causes attenuation and at the same time affects the bass output negatively.

The solution to the problem is an independent amplifier for the headphones such as the circuit shown in diagram 5.0. This amplifier is connected to the output of the final amplifier through the potentiometer P1. If a stereo headphone is used, this potentiometer must be replaced with a stereo type. Furthermore, the entire circuit must be duplicated for the second channel.

The headphone amplifier delivers an output of around 1 watt. Use a power supply rated at 350 mA. The amplifier gain is dependent on the resistors R4 and R6. The values shown in the circuit gives a gain of 11. The voltage at the junction of R13 and R14 must be set at 50% of the power supply. This can be set through P2. The standby current through the final transistors is about 50 ... 110 mA.

6 AUDIO MIXER CIRCUIT

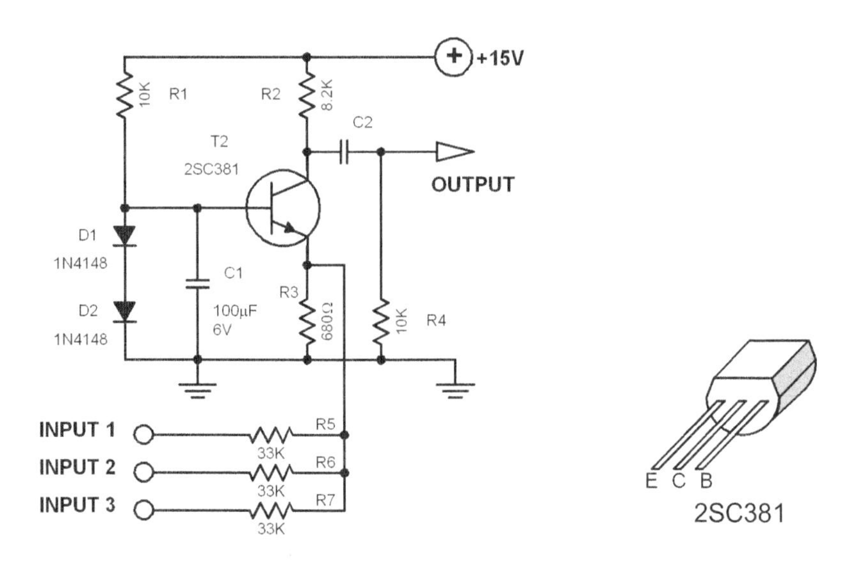

Diagram 6.0 Audio Mixer Circuit

In an amplifier circuit with base driven transistor and with its emitter being current controlled, most of the driving current flows through the collector away. Using the values in the circuit shown in diagram 6.0, the collector current will be about 1 mA. At 15 volts power supply, the input resistors should be 33K. Additional input lines can be connected to the emitter line. Each added input must be series limited by the 33K resistor.

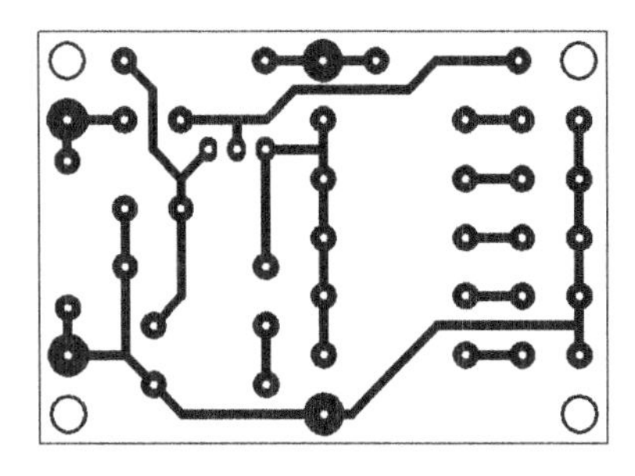

Figure 6.0
Printed Circuit Layout

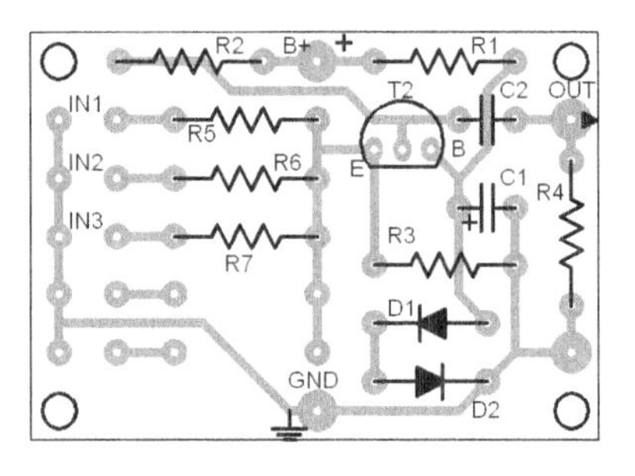

Figure 6.1
Parts Placement Layout

7 SUBWOOFER FILTER

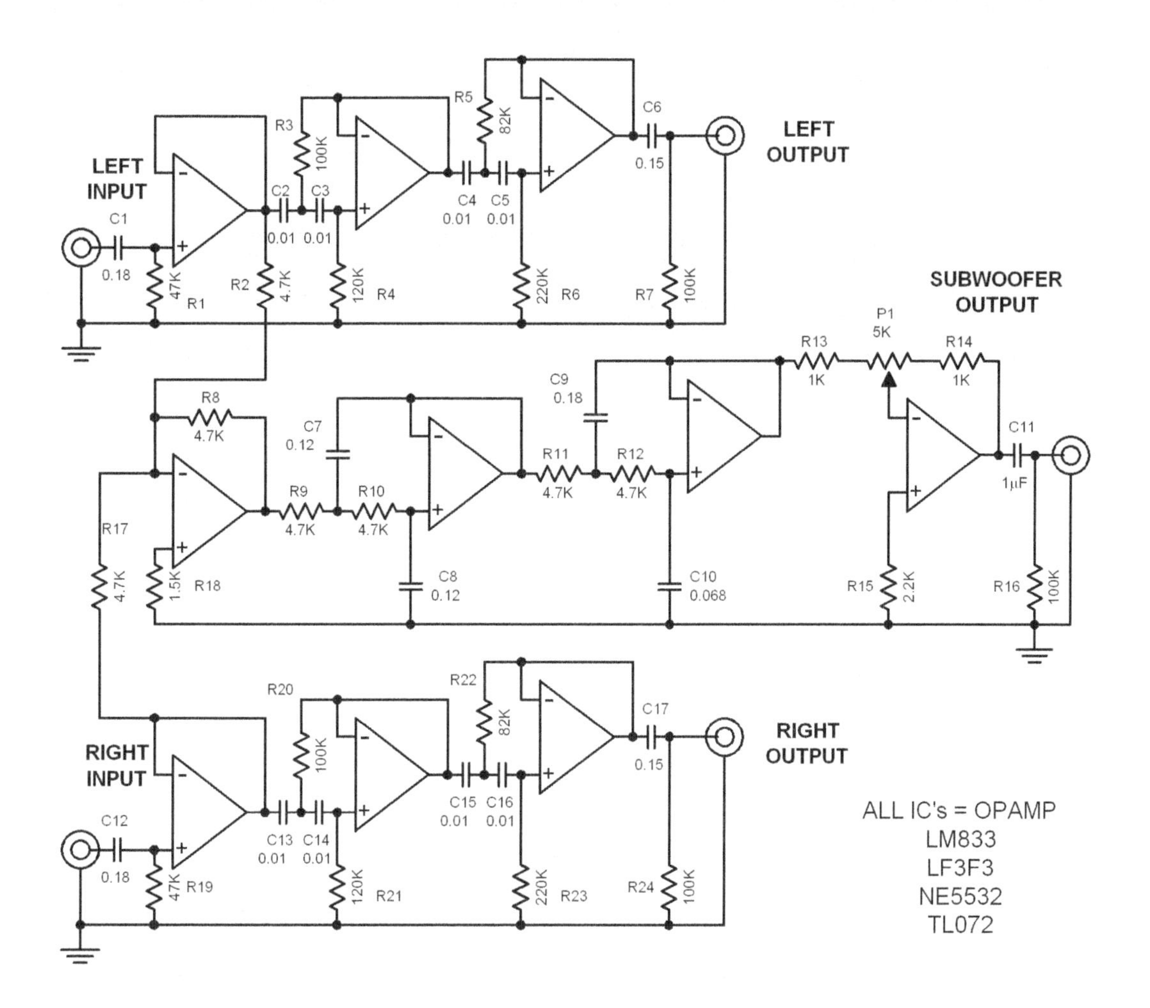

Diagram 7.0 Subwoofer Filter

If you are interested in experimenting with audio circuits in subwoofer range, this circuit is for you. In subwoofer range, all audio frequencies below 200 Hz can be can be fed to a single speaker box (mono) since the human directional perception of sound diminishes at this frequency range. In short, you don't need to use (or buy) two bass speakers for both of the stereo channels. You use only one bass speaker. The normal stereo signals above 200 Hz can be fed to two satellite speaker boxes.

The circuit shown in diagram 7.0 is basically an active filter. It is a 24 dB octave filter with a Bessel character and cutoff frequency of 200 Hz.

How does the circuit work: Opamps A1 and A2 buffer the signals coming from right and left channels. Opamp combinations A3/A4 and A9/A10 function as the highpass filters for both stereo channels. The outputs are then connected to the final amplifiers of the satellite boxes. Signals from both left and right channels are fed to the opamp A5. Opamps A6/A7 function as the lowpass filter. Opamp A8 works as the output amplifier for the subwoofer signal. The signal level can be balanced between the subwoofer and the satellite lines.

The power needed for this circuit must have a symmetrical output. The opamps can have either JFET or bipolar inputs.

8 3- WAY AUDIO EQUALIZER

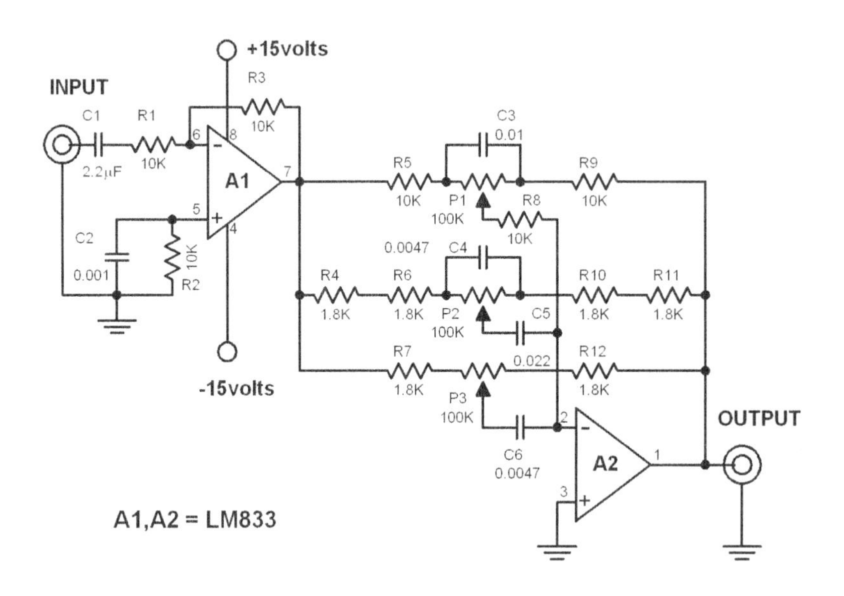

Diagram 8.0 3-Way Audio Equalizer

The circuit is an active filter network for bass, mid and high audio ranges. It is designed around the LM833 an opamp from National Semiconductors. This opamp has the following characteristics: very low noise figure (4.5 nV/sqr(Hz)), wide bandwith (15 MHz at Vu = 1), and relatively high slew rate (7 V/μs).

The output is designed to be DC coupled, however due to slight DC variations through the 100K potentiometers at the feedback lines of the opamp A2, a coupling capacitor might be needed.

Technical Specifications for 3-way Audio Equalizer

The cutoff frequencies:
bass range = 200 Hz
high range = 2 kHz

The midrange is a bandpass network with a center frequency of 1 kHz. The maximum equalizer range is about 16 dB. In the middle position of potentiometer, the noise attenuation is about 90 dB with a banwidth of 1 MHz and a gain of 0 dB (zero amplification). The gain can be changed through R2 using the following formula: Vu = R2/R1.

9 MUSIC PROCESSOR

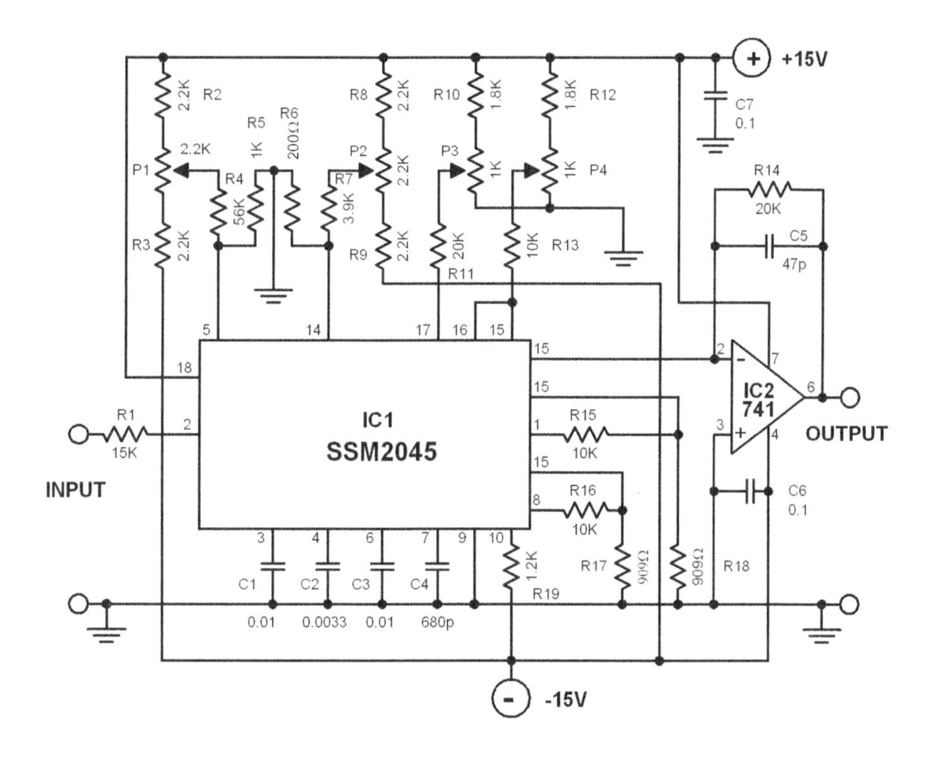

Diagram 9.0 Music Processor

This electronic circuit features the SSM2045 IC which was developed specially for electronic music applications. The IC features a universal application. The circuit is basically configured as a low pass filter with a DC voltage control for gain. The input signal is set to a working level of 150mVp-p through the resistor R1.

The filter has two buffered outputs: the 2-pole output at pin 1 and 4-pole output at pin 8. Internally, these outputs are connected to two voltage-controlled-amplifiers (VCA). The resistors R15 and R16 are connected to these outputs to achieve optimum offset and control voltage suppression. Potentiometer P4 is the volume control. The current that flows to the pins 15 and 16 should not go beyond the maximum of 250 μA. The balance of the two VCAs and the entire filter is being controlled by a voltage range of -250mV to +250mV at pin 14. This voltage can be set by potentiometer P2.

The input can be driven with source impedances up to a maximum of 200 ohms. With an input level of 0dBm, the VCA weakens by 6 dB. The bias current needed at pin 17 (Q input) is between 120 μA and 185 μA. The cutoff frequency can be shifted between 20 Hz and 20 kHz with a variable voltage at pin 5. This can be varied though the potentiometer P1. The capacitor values were selected to give the filter its Butterworth characteristics.

The output current of the SSM2045 IC is converted to a voltage output by the 741 opamp (IC2). Any subsequent circuit must be DC decoupled from IC2.

The noise-voltage ratio is about 80 dB.

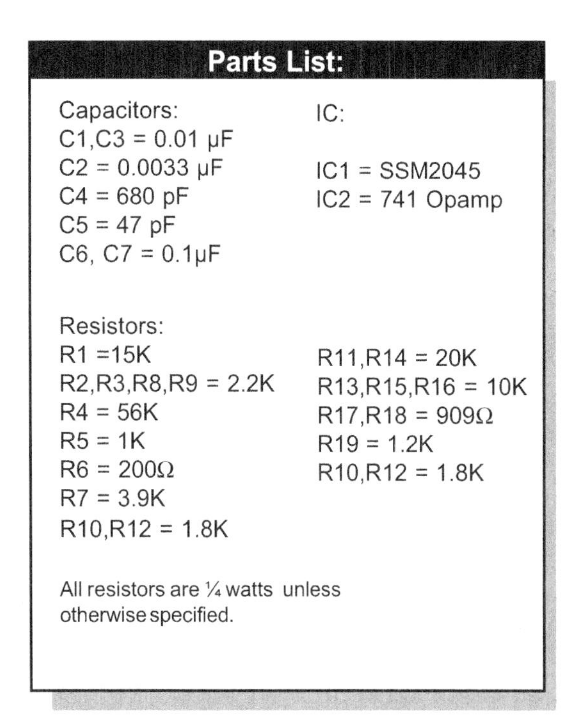

Parts List:

Capacitors:
C1,C3 = 0.01 μF
C2 = 0.0033 μF
C4 = 680 pF
C5 = 47 pF
C6, C7 = 0.1μF

IC:
IC1 = SSM2045
IC2 = 741 Opamp

Resistors:
R1 =15K
R2,R3,R8,R9 = 2.2K
R4 = 56K
R5 = 1K
R6 = 200Ω
R7 = 3.9K
R10,R12 = 1.8K
R11,R14 = 20K
R13,R15,R16 = 10K
R17,R18 = 909Ω
R19 = 1.2K
R10,R12 = 1.8K

All resistors are ¼ watts unless otherwise specified.

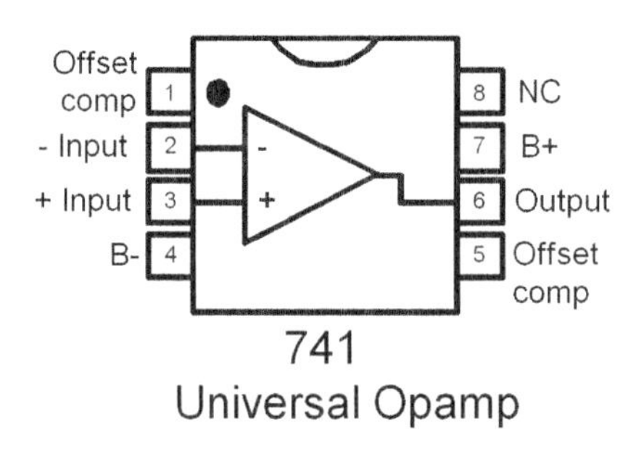

741
Universal Opamp

10 AUDIO FREQUENCY GENERATOR

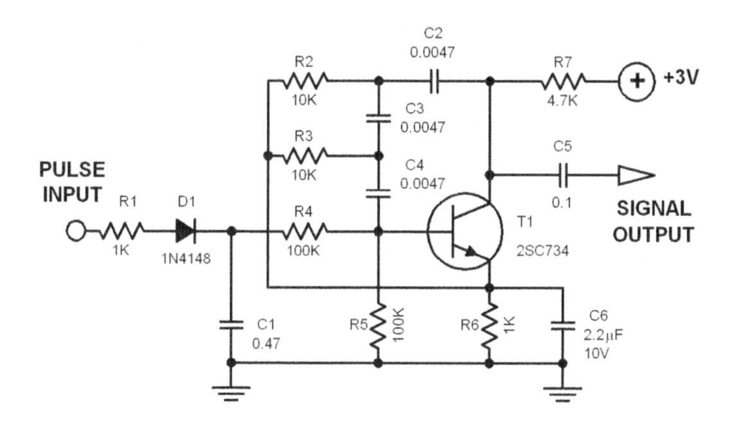

Diagram 10.0 Audio Frequency Generator

This circuit is a triggered signal generator. When a positive pulse of about 6 volts (minimum) is fed to the circuit's input, a modulated audio signal comes out of the output. The signal pattern is similar to a bird's chirp. The pulse width of the trigger signal must be a minimum of 2.5 milliseconds. The voltage supply is between 9 volts and 20 volts. The circuit consumes about 2 mA or less. If the circuit is applied as a morse code signal generator, replace C1 with a 0.1 µF capacitor.

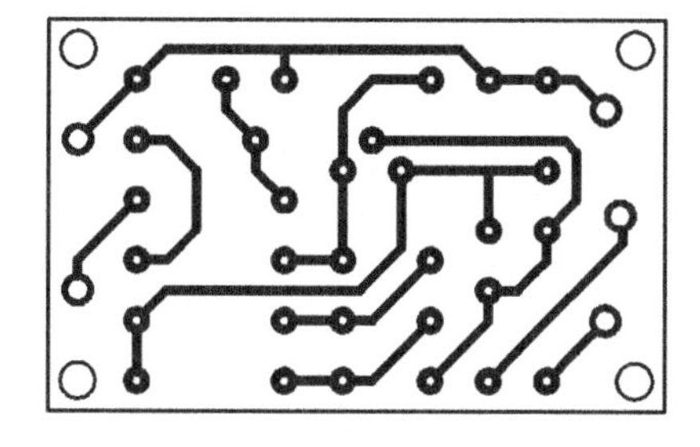

Figure 10.0
Printed Circuit Layout

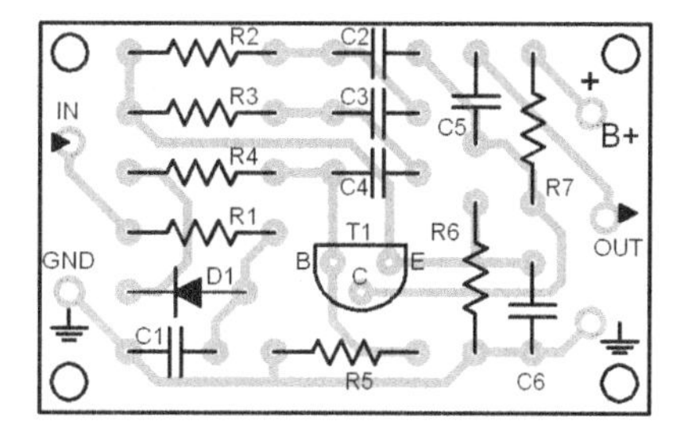

Figure 10.1
Parts Placement Layout

11 DC ELECTRONIC FUSE

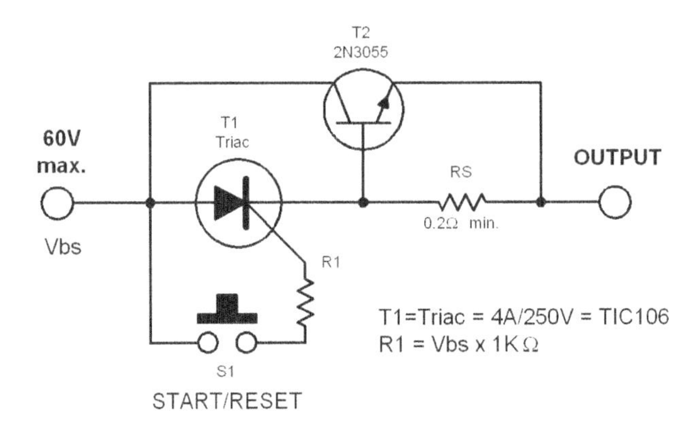

Diagram 11.0 DC Electronic Fuse

This electronic fuse never needs to be ''replaced". It can be repaired with just a single press of the start/reset button S1. Once S1 is pressed, the thyristor T1 triggers and the current flows to the consumer load through T1 and resistor RS. Even after releasing the start button, the current continue to flow as long as the current's value does not sink below a certain level.

The current flowing through the thyristor T1 will sink below the holding level when the current is rerouted through the transistor T2. Transistor T2 and resistor RS are built into the circuit for this purpose. If the voltage drop at RS exceeds above the base-emitter-diode trigger voltage of the transistor T2, the transistor conducts thereby bypassing the thyristor. The resistance value of RS must be at least 0.2W. It must be dimensioned so that the product of RS multiplied by the „fuse" current value equals to 0.7 volts.

Once the transistor T2 bypasses the thyristor, the current flowing through the thyristor sinks below the „holding" level and the thyristor shuts off. This in turn causes the voltage drop at resistor RS to sink below the base-emitter trigger voltage of T2 and the transistor shuts off. The end result is the shutting off of the whole circuit. The „fuse" circuit can be reactivated by pressing the start/reset button.

The value of resistor R1 is dependent on the supply voltage. Multiply the supply voltage with 1 kW to get the value of R1. Connect the circuit to the PLUS line of the consumer load.
The voltage drop at the circuit is less than 1 volt.

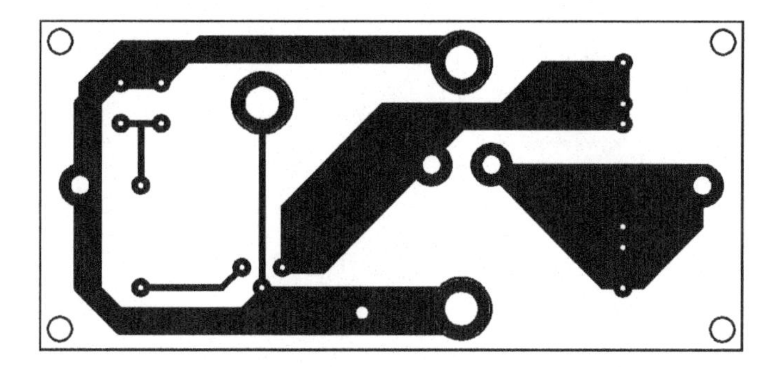

Figure 11.0 Printed circuit layout

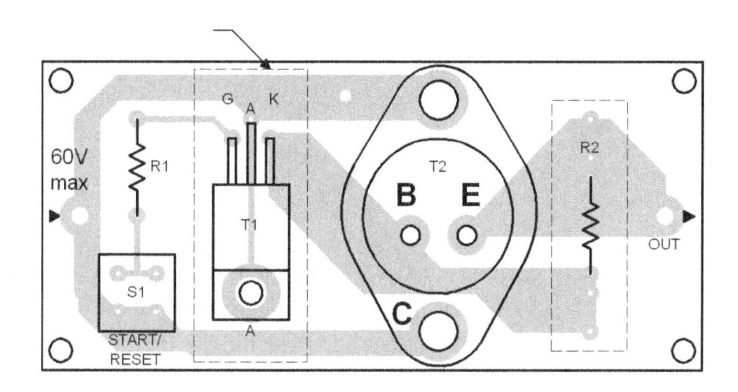

Figure 11.1 Parts placement layout

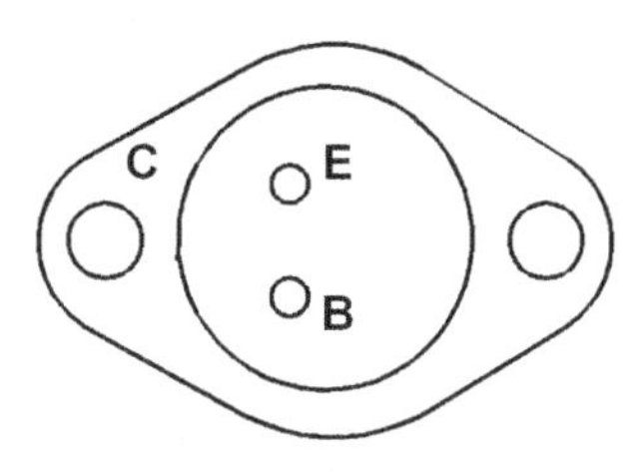

Bottom view

2N3055

12 ISOLATING AMPLIFIER

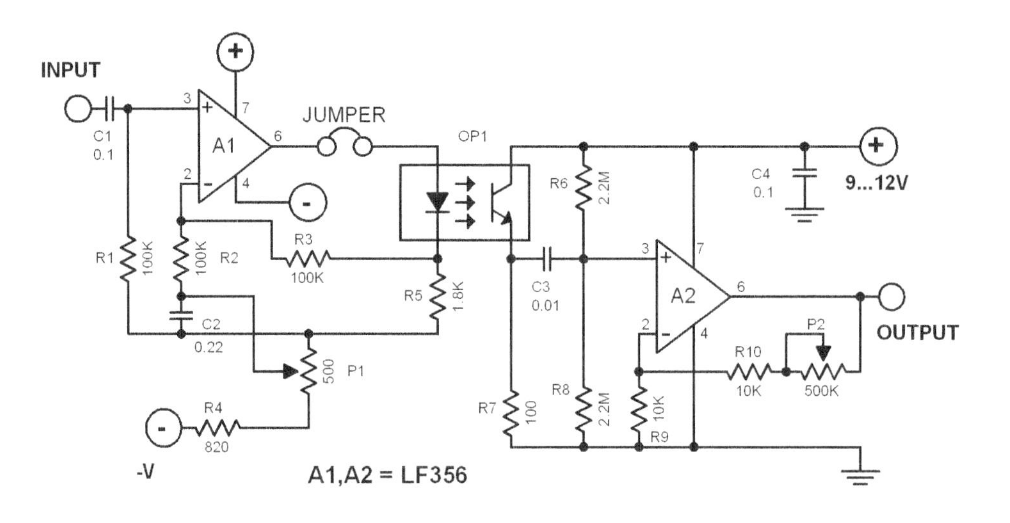

Diagram 12.0 Isolating Amplifier

This isolating amplifier has a bandwidth of 40 Hz up to 40 kHz. With an effective input voltage of up to 70 mVolts at 1 kHz the distortion stays at under 1%. The circuit satisfies the isolation requirements of the safety class 2. The current consumption of each circuit part stays below 10 milliamperes. The core element of the circuit is the optocoupler OP1. It functions as a galvanic separator between the primary and the secondary parts of the circuit. The LED part of the optocoupler is driven by the A1 opamp (half of LF356). The feedback resistor R3 is connected after the LED to keep distortions as low as possible. The working level of the LED is set through P1. During calibration it is important to find a compromise between least distortion and lowest current consumption possible. The distortion factor tends to increase rapidly when the input AC is too high even when the LED is calibrated properly. To avoid this problem, the AC current through the LED must not go over 10% of the LED DC current (at about 70 mV input voltage).

Calibration of the primary circuit is done by measuring the LED current. To do it, remove the JUMPER and replace it with an ampere meter. Set P1 until you measure 1 mA without an input signal.

The signal received through the phototransistor part of the OP1 is amplified by the opamp A2 (second half of LF356). Amplifier gain is adjusted through P2. The circuit needs separate (galvanically isolated) power supplies for each part. Take note that the primary part (the input part) needs a symmetrical supply of 9...12 volts.

13 BLINKER CIRCUIT

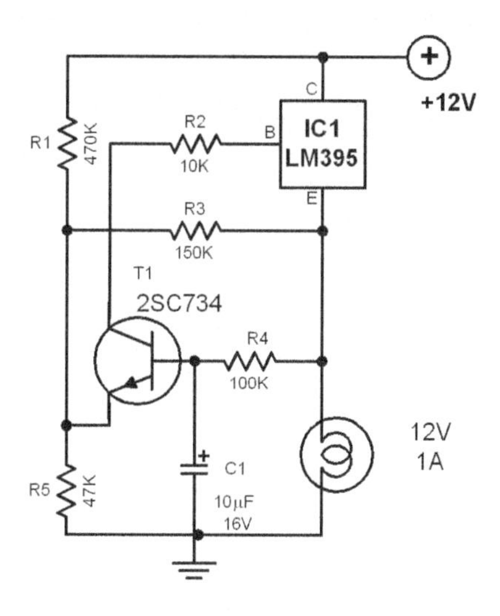

Diagram 13.0 Blinker Circuit

The blinker circuit is made using the LM395 IC. This IC is a short-circuit proof power transistor with special characteristics. This is sometimes called a „super transistor". It is used in this circuit as an alarm blinker for cars.

The blink frequency is determined by the R4/C1 combination which is dimensioned to give approximately one blink per second. To achieve lower frequencies, increase the value of C1. To achieve faster blink frequency, decrease the value of C1. The circuit can drive 12 volt lamps. The maximum power delivered is 12 watts. The power transistor T1 must be heatsinked. Two PCB designs are available: one for the TO-3 package and one for the TO-220 package.

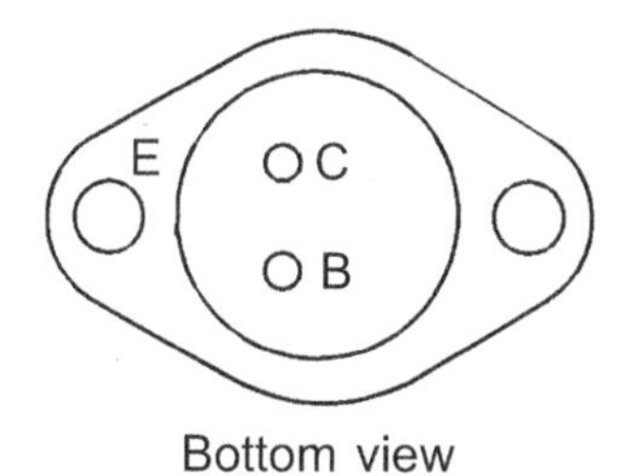

B E C

LM395

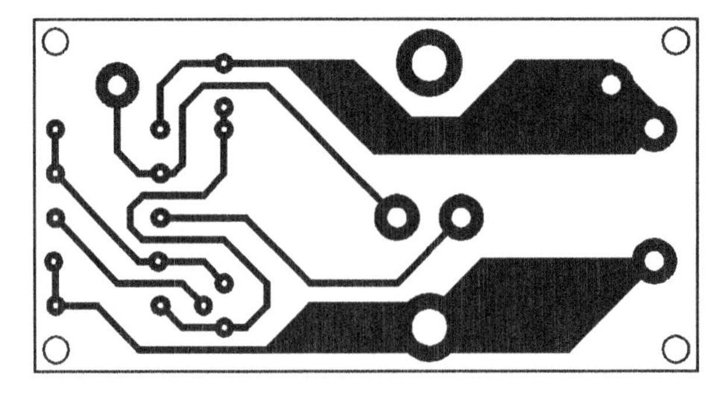

Figure 13.0
Printed Circuit Layout for TO-3 package

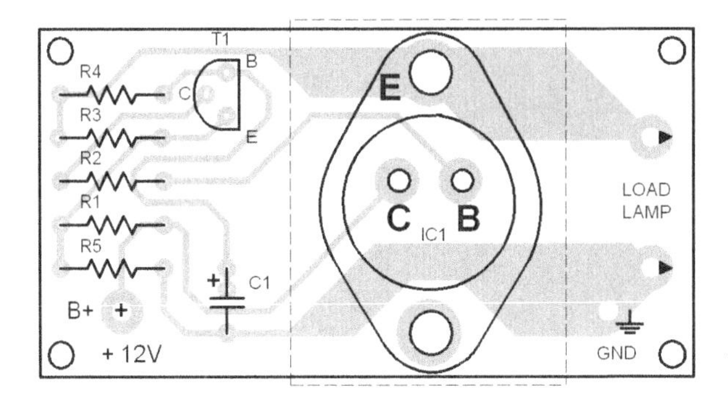

Figure 13.1
Parts Placement Layout for TO-3 package

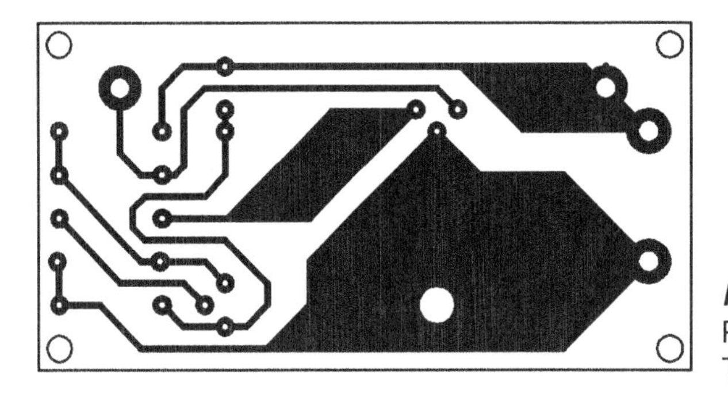

Figure 13.2
Printed Circuit Layout for TO-220 package

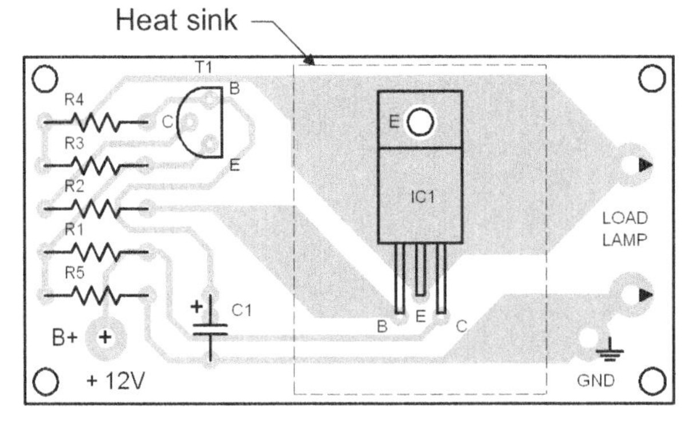

Figure 13.3
Parts Placement Layout for TO-220 package

14 CRYSTAL FILTER

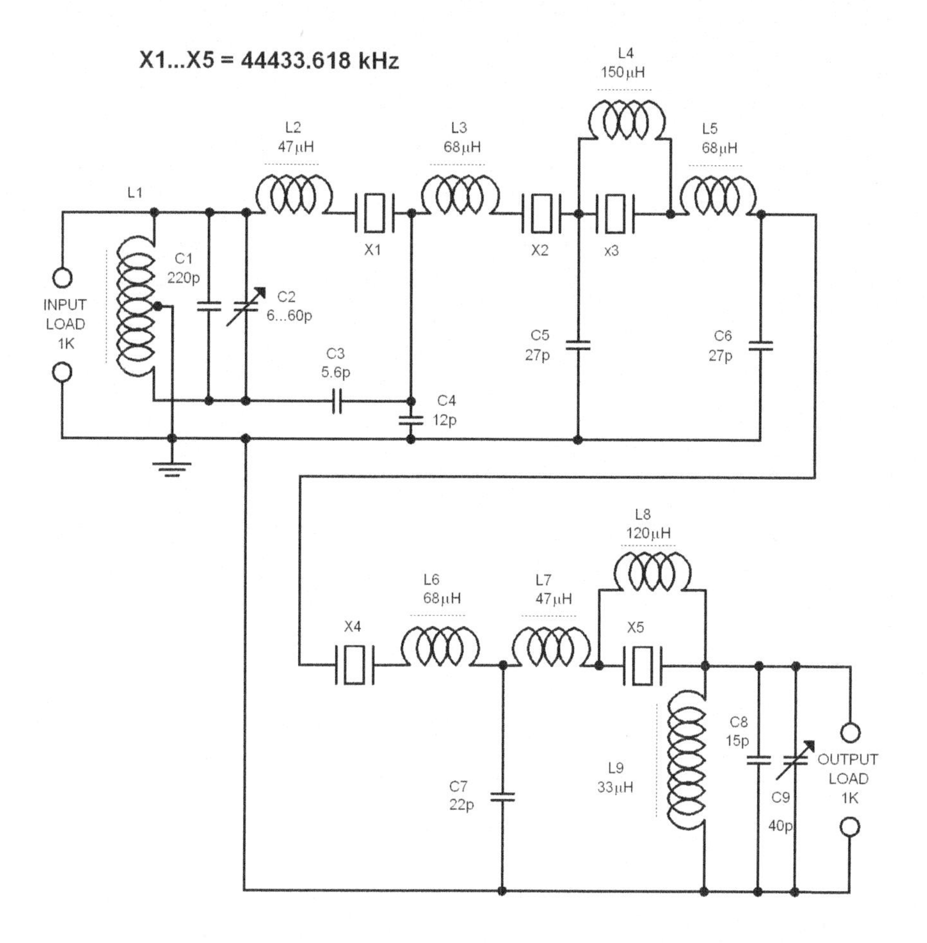

Diagram 14.0 Crystal Filter

In constructing CB band receivers, the biggest problem is most often the selectivity of the circuit. This is due to the fact that the channel separation is usually between 9 and 10 MHz only, a very narrow separation. To avoid cross channel interference, a receiver must have a very good filter network. The featured circuit is one such filter network. Its special feature is its use of readily available crystals commonly used in television sets. These crystals are fairly cheap too. The only downside of the crystal is that it is bound to the resonant frequency of 4.433618 MHz. For the mentioned application however, this frequency is good enough.

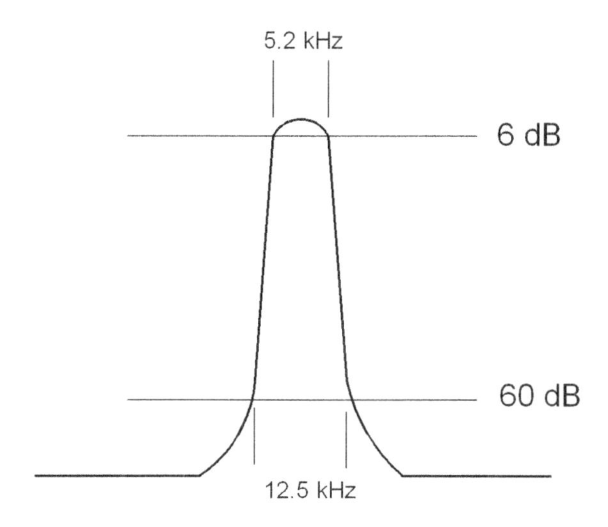

Graph 14.0 Bandpass characteristics

You can see from the diagram that the circuit is basically a ladder filter composed of 5 crystals. This crystal configuration normally results to an irregular feedthrough characteristics. The circuit however was dimensioned carefully to achieve a symmetrical feedthrough curve. As shown on the graph 14.0, the 6 dB bandwidth is about 5 kHz and the 60 dB point is at 12 kHz. All coils are readily available from supply stores except L1. Coil L1 is made with 15 bifilar turns of 0.4mm copper wire in a Amidon T50-2 ferrite core.

It is important to shield each filter stage from each other using appropriate faraday shields. Electrically connect (solder) all crystal metal housing to the circuit ground too.

A good filter network can enhance the quality of the received radio information

15 ELECTRONIC ORGAN

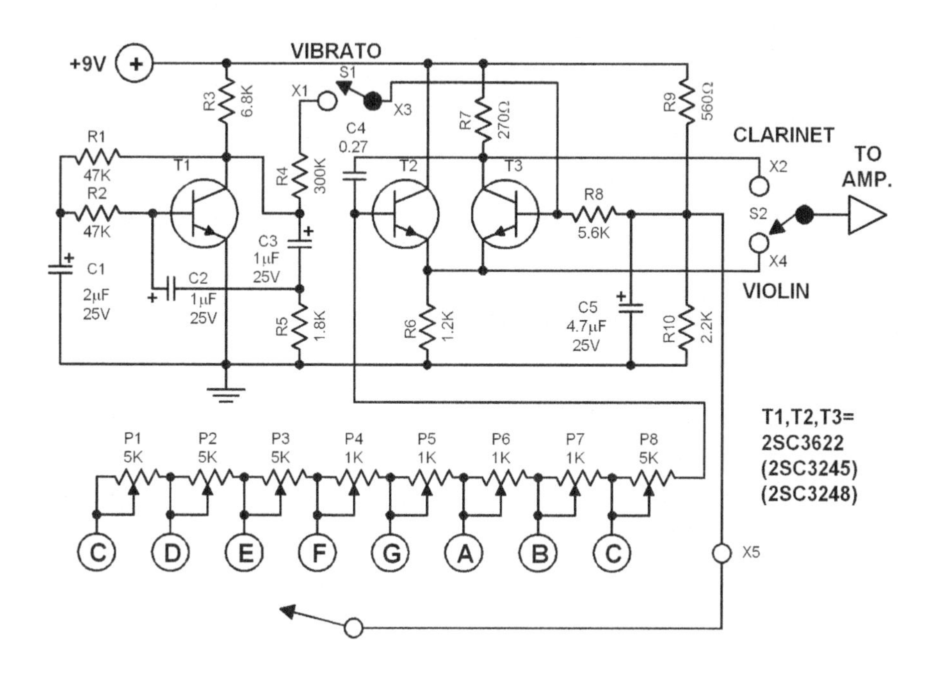

Diagram 15.0 Electronic Organ

This electronic organ is very simple to construct and can provide hours of enjoyment particularly for children. The circuit is basically an emitter-coupled oscillator composed of T2 and T3. An squarewave voltage can be sampled from the collector of T3(X2). This signal gives a clarinet character to the tone . Without the squarewave signal, the sound produced by the emitters of T2 and T3(X4) has a violin character.

An additional vibrato signal can be added to this basic sound through switch S1. The frequency of the vibrato is approximately 6 Hz. Its amplitude is determined by the resistor R4. The value of R4 can vary from 100 up to 300K. Try experimenting with different values.

The keys can be made of either metal plates or etched printed circuit. The trimmers P1 up tp P8 adjust the pitch of each tone. The tones can be drastically changed by changing the value of C4.

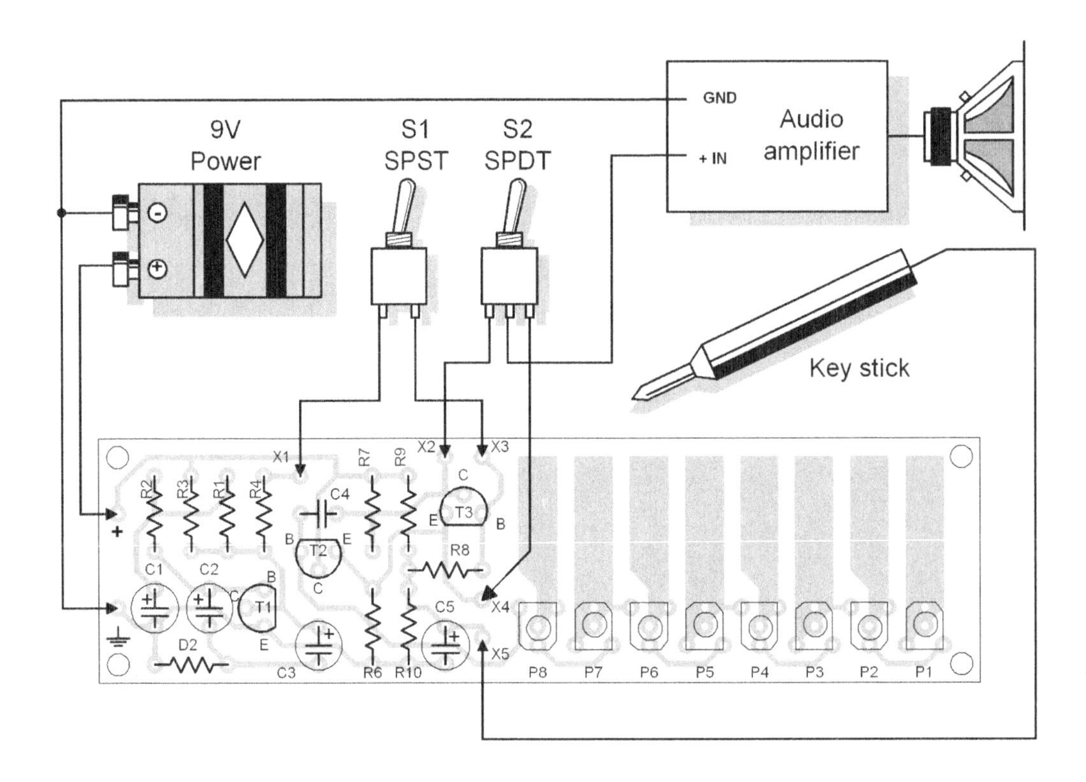

Figure 15.0 Parts Placement Layout for the Electronic Organ

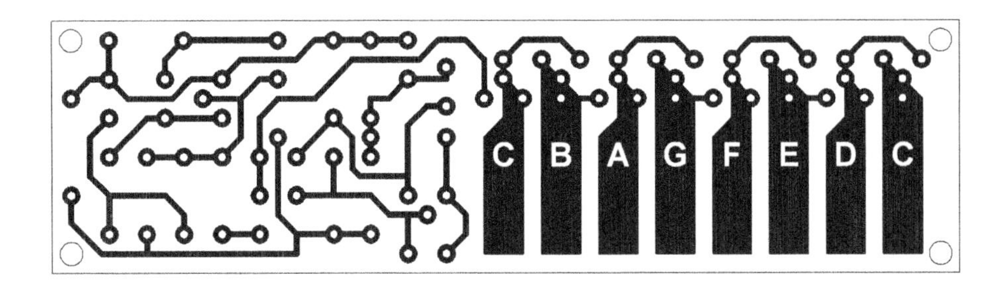

Figure 15.1 Printed Circuit Layout for the Electronic Organ

16 HIFI STEREO PREAMP

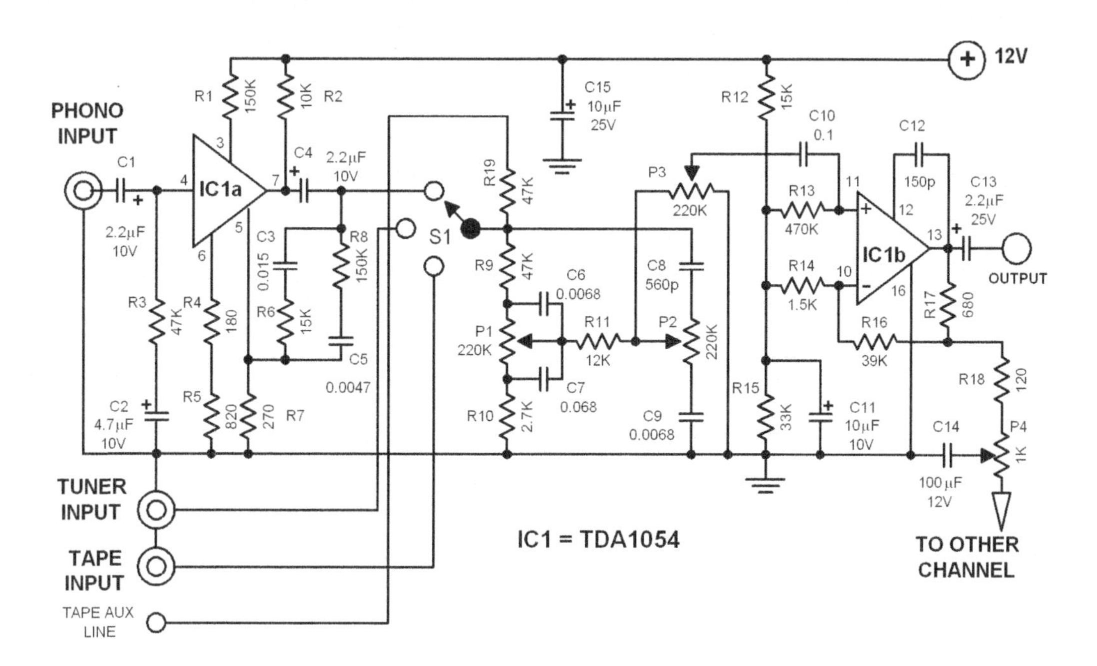

Diagram 16.0 HIFI Stereo Preamplifier

There is an abundance of ready to use final power amplifiers in electronic stores nowadays. Many of these devices have great electrical data such as distortion factor, power output and frequency bandwidths. However, no matter how good these data are, these power amplifiers are still of no good use without a good preamp stage.

Lately, a new breed of preamp IC's came to the market. These IC's allow the hobbyist to build low noise preamp circuits with little or no problem. One of these IC's is the TDA1054 from SGS. It is a 16-pin DIL package and integrates two separate preamp circuits. The first half of the circuit (IC1a) is a preamp with an input sensitivity of 3 millivolts. It has a frequency correction composed of R6,R8, C3 and C5. The bass signal coming from the phono input is amplified while the high signal is attenuated.

Switch S1 allows the selection of the input signal source. The potentiometers P1 and P2 are part of a double potentiometer. They control the bass and high tones. There is no risk of overdrive in this circuit due to the passive nature of the sound control. Potentiometer P3 controls the volume of the signal fed to the second part of the circuit (IC1b) which functions as an operation amplifier. P4 controls the balance between the left and right channels (one channel not shown on the diagram). At the middle setting of P4, both channels have a gain of 24.

If P4 is set to one extreme end, the gain difference between the two channels is about 12 dB.

Diagram 16.1 shows the voltage regulator circuit for the preamp. It uses the 7812 IC and supplies 12 volts.

Input sensitivity at a frequency of 1 kHz and output of 775 mVeff:

Phono = 3mV/50k
Tuner = 220mV/50k
Tape = 220mV/50k

Maximum output voltage = 2.5 eff.
Balance range = 12 dB

Sound adjustment:
low = +/- 13 dB (100 Hz)
high = +/- 13 dB (100 Hz)

Distortion = <0.05% (f = 1 kHz, output voltage = 775 mVeff)

Bandwidth = 20 Hz ... 24 kHz

Signal to noise ratio = >65 dB (output voltage = 775mVeff)

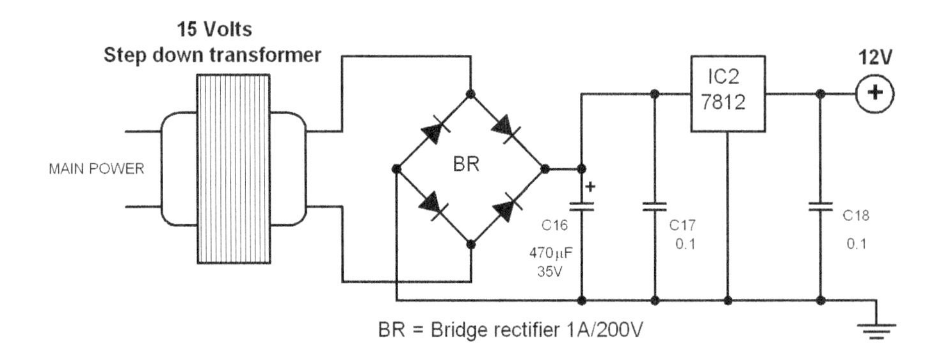

Diagram 16.1 Regulated power supply for the HIFI stereo preamplifier

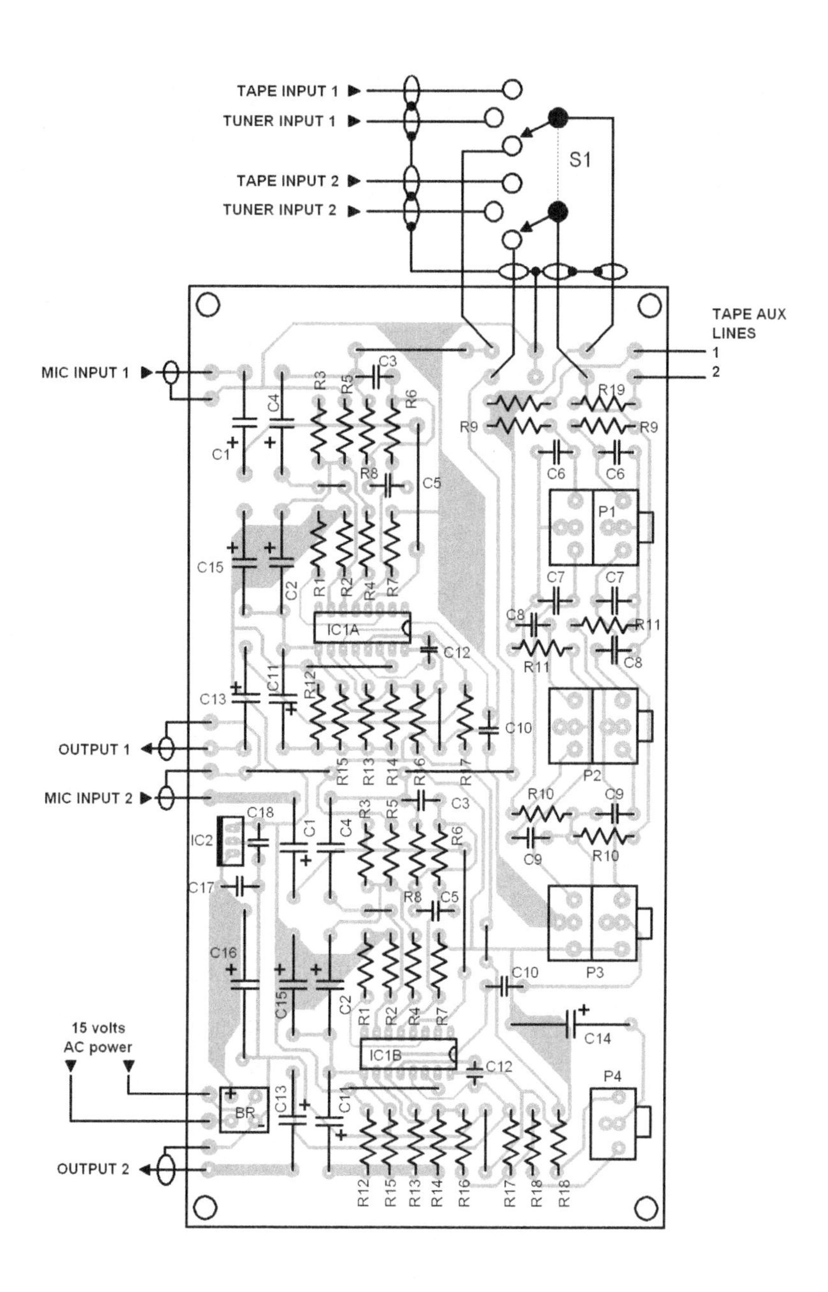

Figure 16.0 Parts Placement Layout for the HIFI Stereo Preamplifier with external wiring guide

Caution! Danger of electrocution!
Extreme shock hazard! You are working with a line voltage of 220 Volts AC.

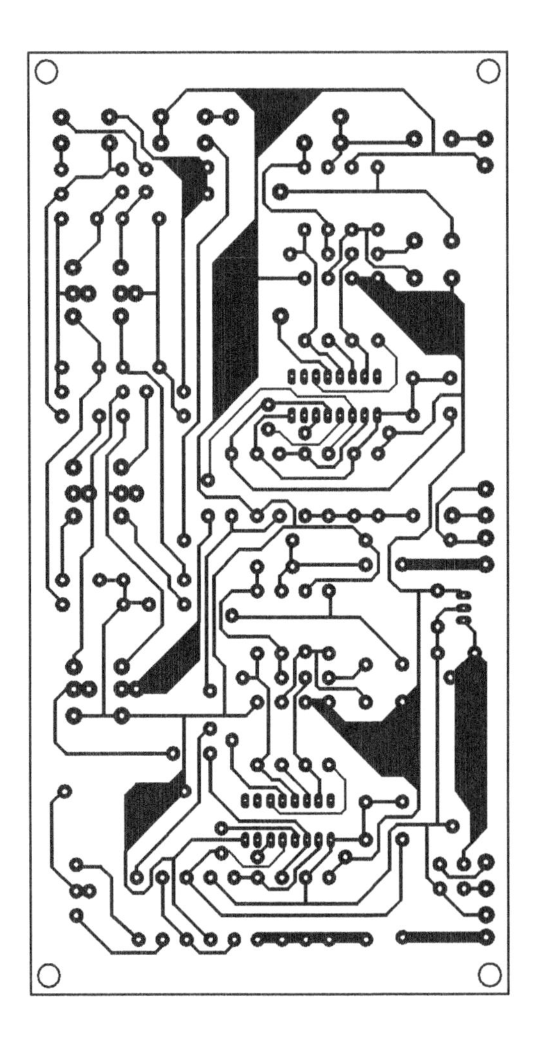

Figure 17.0 Printed Circuit Layout for the HIFI Stereo Preamplifier

17 FM STEREO NOISE SUPPRESSOR

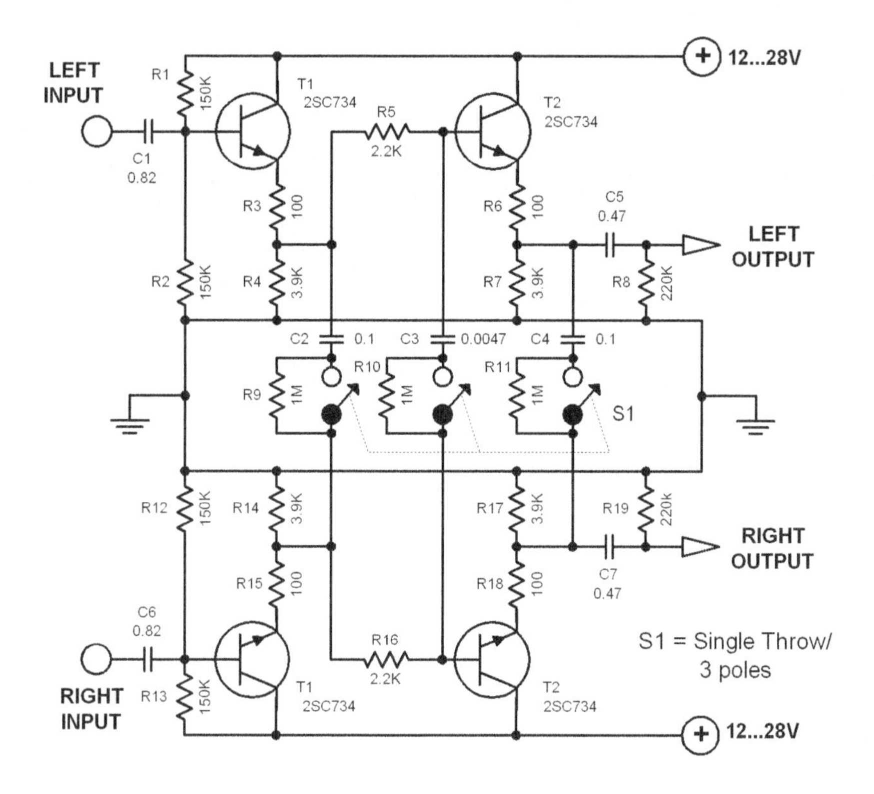

Diagram 17.0 FM Stereo Noise Suppressor

Sometimes reception of far FM stations is bad but improves when you switch the radio receiver to mono. The cause of this effect is that in stereo reception the biggest part of the noise in the right channel is in opposite phase with the noise in the left channel. Once the two channels are switched together (mono), these noise signals cancel each other. The result is a clean and relatively noise free mono signal. To use this effect in improving the signal but keeping the stereo character, the two channel should be switched together only in higher audio frequencies. This is done with the help of the above featured circuit.

The circuit is built in two identical channels each composed of two emitter-follower circuits. The switch S1 can connect the two channels together at three points in the circuit. The circuit functions as a bridge between the left and right channel for opposing signal components above 8 kHz. These signals get „short-circuited“ cancelling the noise. Otherwise, all other frequencies get through to the output via their respective channels.

If you want to lower the „short-circuit“ frequency to 4 kHz, just double the values of C2, C3, and C4.

18 TEMPERATURE MONITOR

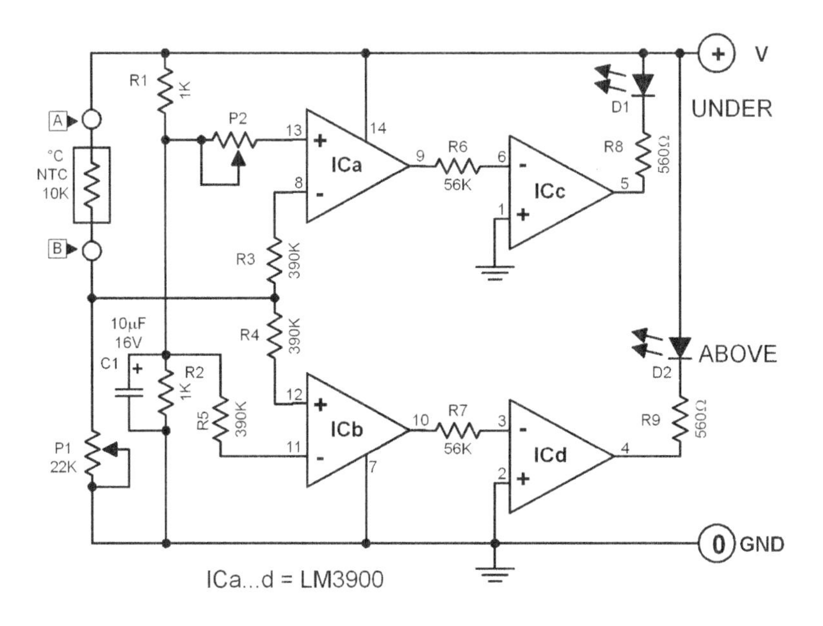

Figure 18.0 Temperature Monitor

Sometimes a continuous analog display of a temperature value being monitored is not necessary. A simple indication whether the temperature value exceeded a maximum level or went below the minimum level is sometimes enough. The circuit here does just that. It indicates the temperature level above +25°C (77°F) or below +20°C (68°F) by lighting one of the two LEDs. The temperature sensor is an NTC resistor coupled to two comparators. When D3 lights, the temperature is above +25°C (77°F). Conversely when D2 lights, the temperature is below +20°C (68°F).

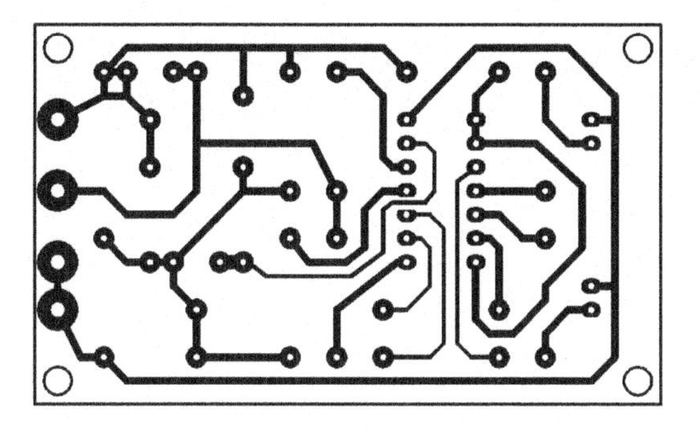

Figure 18.0 Printed Circuit Layout

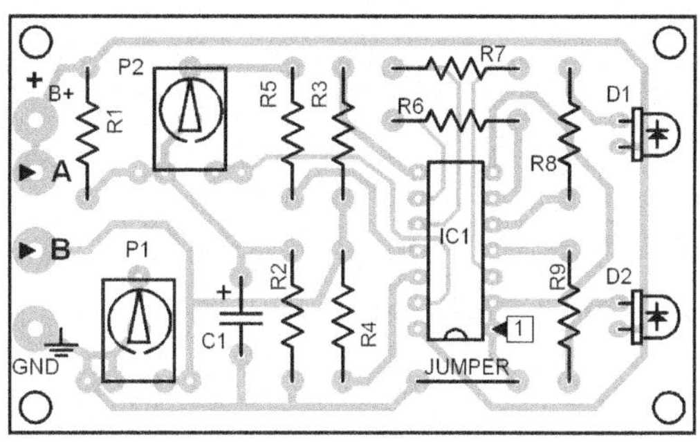

Figure 18.1 Parts Placement Layout

Calibration:

Place the NTC resistor in cold water. Slowly heat the water until the desired maximum temperature level is reached (use a thermometer), then adjust P1 until D2 lights up.

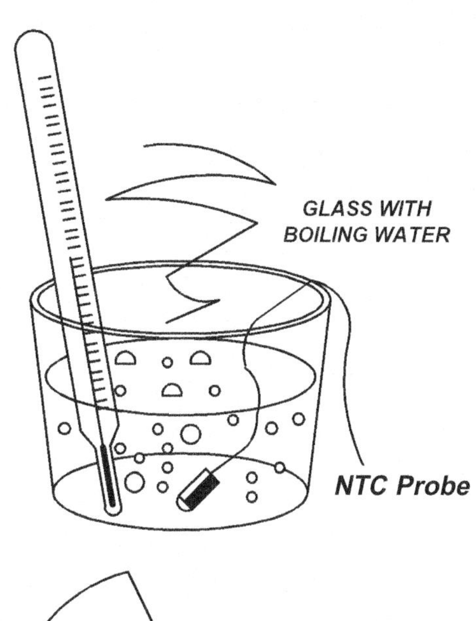

To set the minimum level, stop heating the water then slowly add cold water to it until the desired lower temperture level is reached as indicated by the thermometer. This time, adjust P2 until D1 lights up.

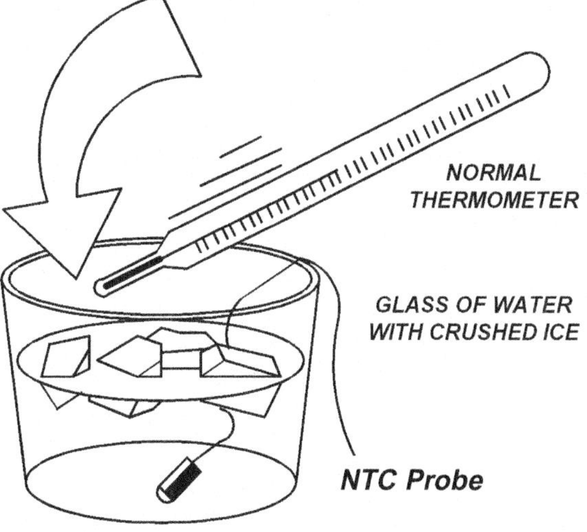

19 SPEAKER BALANCE

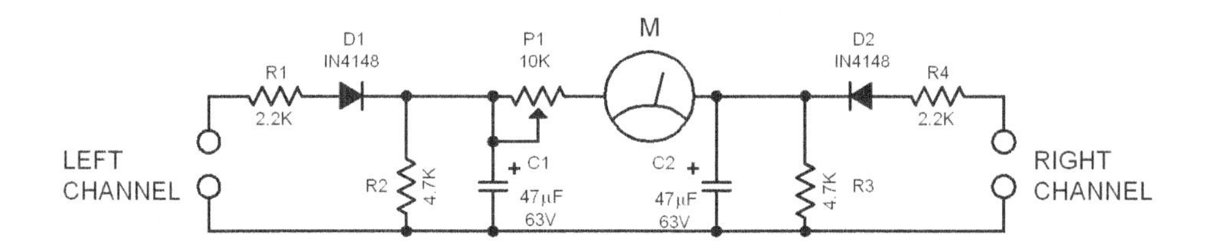

Figure 19.0 Speaker Balance

When using a stereo amplifier, there are many possible mechanical problems that may influence the amplifier's output performance. One would wish to avoid them but they happen more often than one thinks. However, the unwanted effects are usually very subtle that they don't even get noticed at once. The usual culprit is the mechanical potentiometers which are not rarely unsync in their stereo resistance pads. This results to unbalanced volume levels on the audio channels. Most good stereo units have a balance control to compensate for this imbalance.

In order to know exactly the imbalance and be able to correct it, one needs the speaker balance indicator circuit featured here. In using the circuit, its left and right input channels are connected to the corresponding speaker outputs of the amplifier. Identical signals are then fed to the amplifier's stereo inputs. The best signal would be a pure sine wave from a signal generator. When a signal of exactly the same amplitude is coming out from both speaker outputs, the meter M of the speaker balance circuit will stay at its zero setting. The meter M is a type with a zero setting at the middle of the scale. This enables one to see at first glance when one channel is louder than the other channel, e.g. if the meter swings to the left, it means that the signal amplitude at the left channel is higher than at the right channel. To correct the imbalance, one has to turn the amplifier's balance control until the meter goes back to the middle of the scale.

Calibrating the speaker balance indicator circuit: Inject a signal in one channel and set the amplifier's volume to maximum. While doing this, turn the potentiometer P1 until the meter swings to end of the scale in the direction of the channel, e.g. left end of scale for left channel and right end of scale for right channel.

20 TOUCH VOLUME CONTROL

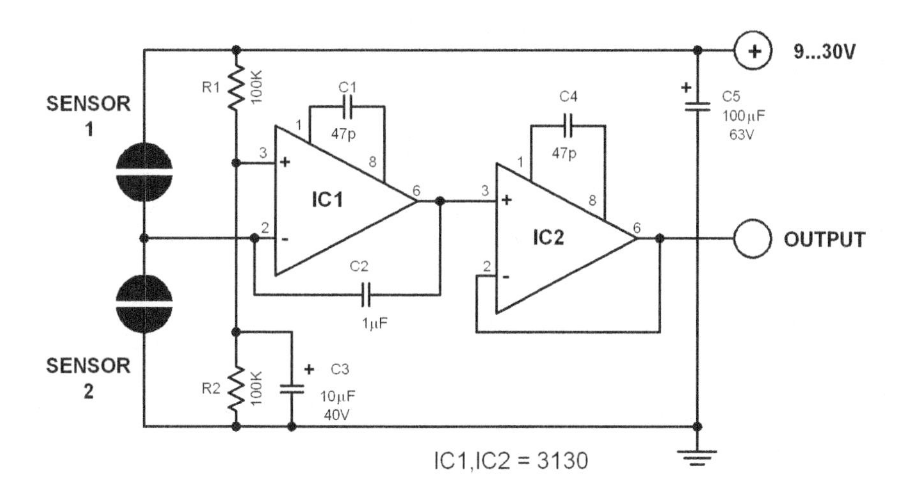

Diagram 20.0 Touch Volume Control

Touch controls are not only used to switch devices on or off. They can also be used to control different functions. One good example is the TV remote control. If it is very important to keep the activated functions for a long period of time, it is always better to use a digital memory system. However, if small drifts in the control status is acceptable, a simple analog design can be used to memorize the status.

The featured electronic circuit is one such analog memory touch control switch. The main function centers mostly on the IC1. It is an opamp configured as an integrator with a high impedance input. If sensor 1 is touched, the capacitor C2 charges through the skin resistance and voltage at the output of IC1 decreases linearly until it reaches zero volt.

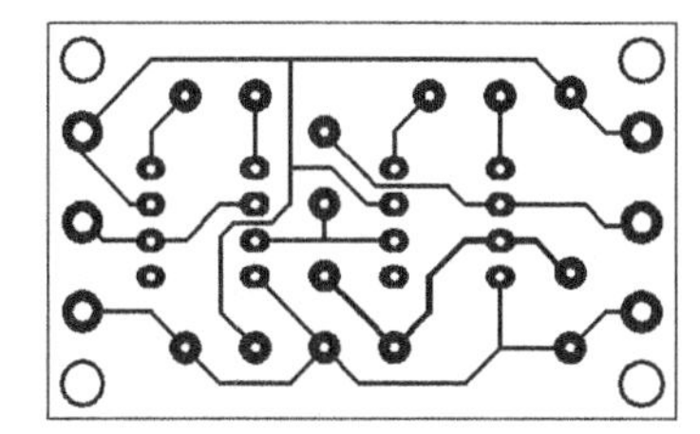

Figure 20.0
Printed Circuit Layout

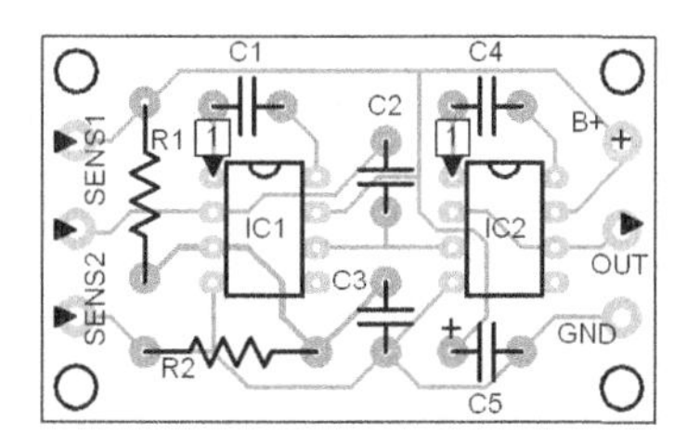

Figure 20.1
Parts Placement Layout

Touching the other sensor (sensor 2) will produce the opposite result: the voltage at the pin 6 of IC1 will increase linearly until it reaches the power supply level. The special function of this circuit is that after moving the finger away from the sensor(s), the output voltage of IC1 stays at that level. This voltage value is memorized by C2. This analog memory however has a problem in long time periods: the voltage value drifts away by 2% per hour due to the unavoidable current leak in the capacitor. To improve this situation, it is highly recommended to put the circuit in a moisture proof box.

This circuit has a wide application range. It can be used in devices where a potentiometer can be controlled through voltage levels. The touch sensors can also be replaced with conventional push button switches. The capacitors C1 and C4 are very important in the circuit: they prevent the IC1 from oscillating. Simultaneously closing both switches will not damage the circuit.

21 1-CHIP 40 WATT AMPLIFIER

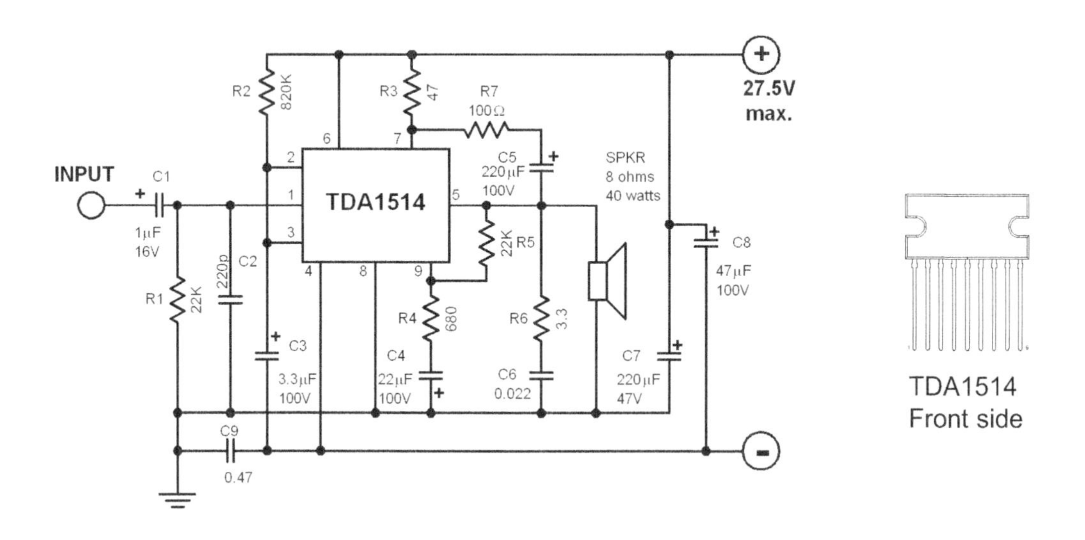

Diagram 21.0 1-Chip 40 Watt Amplifier

This is a compact, easy to build amplifier that uses one IC only but delivers 40 watts of audio power! It is ideal for amplifying audio from your mobile CD player or iPod. The chip being used here is the TDA1514 originally brought to the market by Phillips/Valvo. The best characteristics of this chip is its high output power and robustness. It is available in a 9-pin SIL plastic package with a metal mount.

Its package has a heat resistance of less than 1.5K/W. This means that the heatsink must have a heat resistance of only 3.8K/W when the chip reaches its maximum power dissipation of 19 watts (at Ub = +/- 27.5 V, Tu = 50°C).

One can see from the diagram that only a handful of passive elements are needed to build the chip into a powerful audio amplifier. The power supply as supplied must be able to deliver a current of at least 3 amperes. The standby current consumption is about 60mA. The supply voltage must *never* go beyond 27.5 volts!

In building the circuit, keep the wires to the power supply and outputs as short as possible. The resistors R4 and R5 set the voltage gain at the feedback which, in this case, is between 20 and 46 dB.

For a single channel amplifier (mono), a 80VA transformer (T1) should be sufficient. If you construct two channels (or stereo) amps, 120VA is recommended. Capacitors Cx and Cy should be at least 4,700mµF rated at 35V. It can be up to 10,000µF. Capacitors twice as large discharge slower giving better peak power potential resulting to better power output. Feel free to increase the capacitance but take note that you may not get much additional benefit for the price involved. Make sure they are connected the right way around too or they will blow and cause injury.

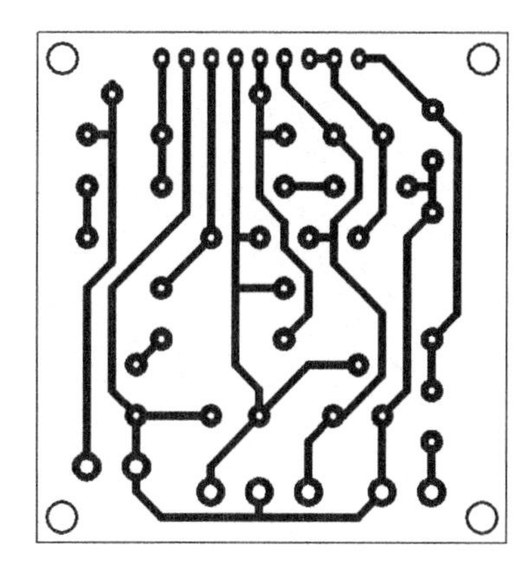

Figure 21.0 Printed Circuit Layout

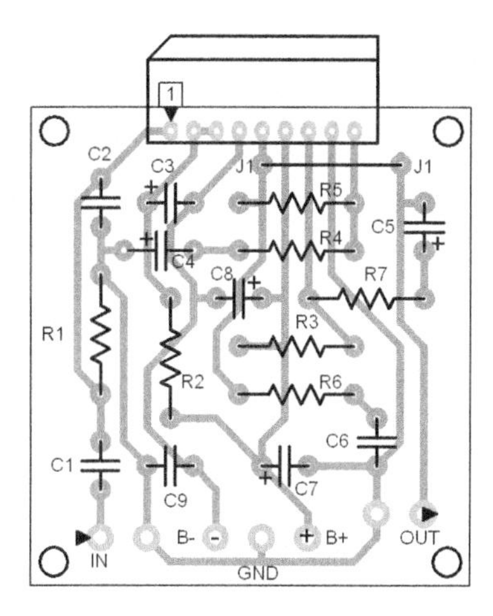

Figure 21.1 Parts Placement Layout

A 0.25mA fuse (F1) should be installed. If using a toroidal transformer, the fuse must be a time delay (slow blow) variant. Be sure to correctly earth the supply and any metal casing around it. The components on the earth and ground connections (Dx,Dy,Rx,Cz) form a loop breaker. This is recommended because it can eliminate those troublesome earth loops. Rx is a 5W or better wirewound resistor. The Cz capacitor must be rated for 250V AC. Do NOT use a 250V DC cap for Cz as it would fail if there ever was a fault causing mains current to flow to earth. Dx and Dy are rated at 250V/1A. If your local rules and regulations prohibit constructing this, omit all these components and connect the earth to ground but *NEVER* disconnect the earth lead... it could save your life or somebody elses!

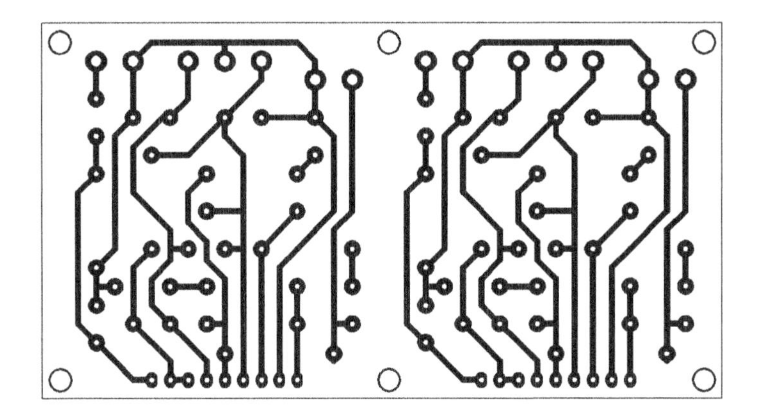

Figure 21.2 Printed Circuit Layout (stereo version)

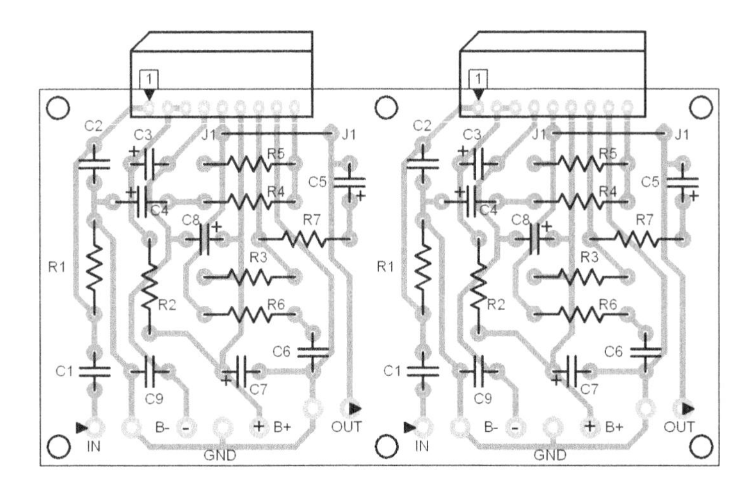

Figure 21.3 Parts Placement Layout (stereo version)

Power Supply:

Constructing the power supply for this amplifier is simple. As shown on diagram 21.1 you need to wire up a 18-0-18 (center tapped) transformer in order to get the recommended +/-25V. Be very careful since this construction involves mains wiring:

MAINS VOLTAGE IS VERY DANGEROUS.
DO NOT WIRE IT UNLESS YOU ARE QUALIFIED.
DEATH OR SERIOUS INJURY MAY RESULT.

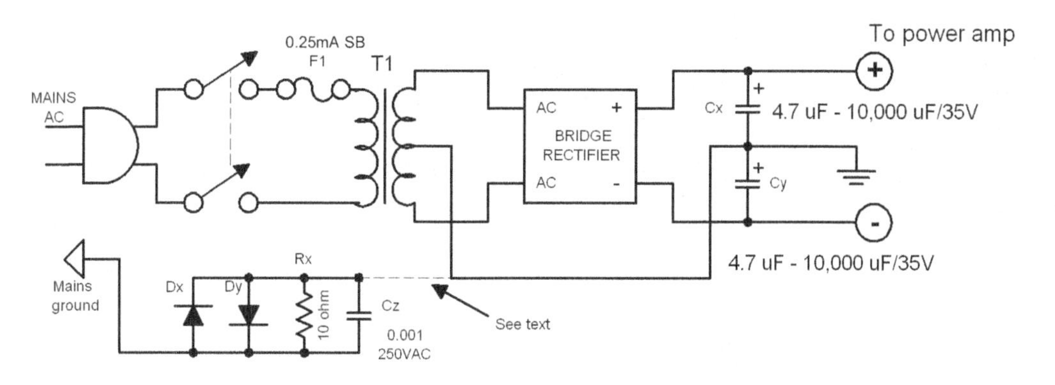

Diagram 21.1 Power supply for 1-chip amplifier

22 HEATSINK MONITOR

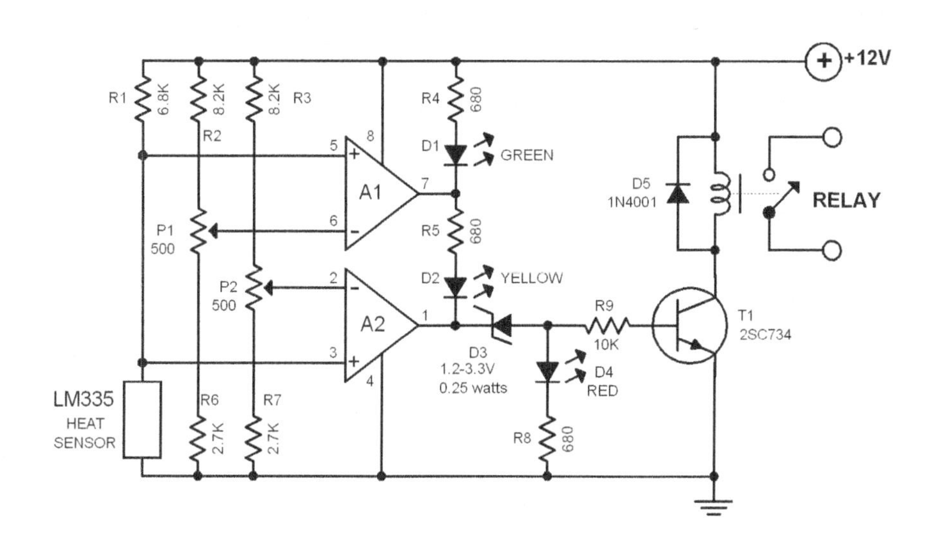

Diagram 22.0 Heatsink Monitor

The heat dissipation from power transistors and diodes in electronic devices are typically absorbed by heatsinks and further dissipated into the environment. The heatsink dimensions are based on the maximum allowable crystal temperature of the semiconductor as published by the manufacturer. However, the actual temperature reached by the heatsink is not clearly known. Many people test the heatsink temperature with their fingers. It seems that as long as it does not burn their fingers, it is accepted to be safe. Obviously, this method is not accurate that is why the electronic circuit featured here was designed.

This heatsink monitor signals via three LEDs when the temperature exceeds two boundary levels. When the heatsink temperature is below 50 ... 60 degrees centigrade (122 ... 140 degrees fahrenheit) the green LED lights. The yellow LED signals that the temperature is within 60 and 70 degrees centigrade (140...158 degrees fahrenheit). The red LED signals that the temperature has reached beyond 70...80 degrees centigrade (158 ...176 degrees fahrenheit). At this high point a relay can be triggered to break the power supply or otherwise protect the electronic device by any other means (e.g. turning on a cooling fan).

As shown in the diagram above the electronic circuit is made of two opamps that are set as window comparators. The input voltage level is set by the thermosensor IC LM335.

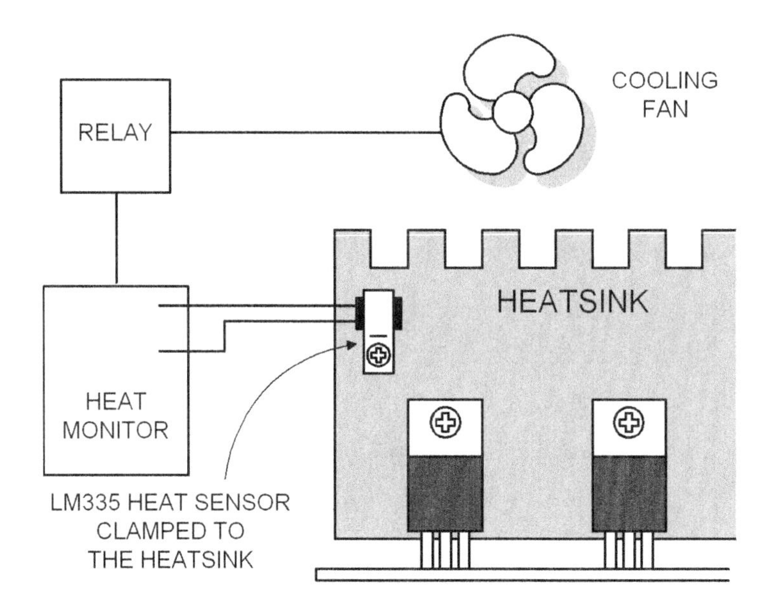

Figure 22.0 External Wiring Diagram. The LM335 heat sensor is clamped to the heatsink being monitored.

This voltage level increases linearly by a factor of 10mV/°centigrade. If this voltage is below the voltage level set by the potentiometers P1 and P2, the opamp outputs are almost at ground potential (0 volts) and the green LED lights up. If this voltage is within the set boundaries, the output of A1 is „high" and the yellow LED will light up. If this voltage exceeds the voltage level set by P2, both A1 and A2 outputs are „high" and the red LED lights up while the yellow LED turns off. At the same time, the load at T1 is reduced to prevent further increase in the heatsink temperature.

To set the desired temperature levels: put the thermosensor LM355 together with a thermometer into a pot of water which is being heated slowly. Turn P1 to its lowest point and turn P2 to its maximum point. The switching point from green LED to yellow LED (50...60°C or 122...140°F) is set with P1. The switching point from yellow LED to red LED is set with P2. Watch the thermometer while the water is being heated and adjust the potentiometers accordingly. Once the desired temperature boundaries are set, attach the thermosensor to the heatsink.

23 X-Y VU DISPLAY

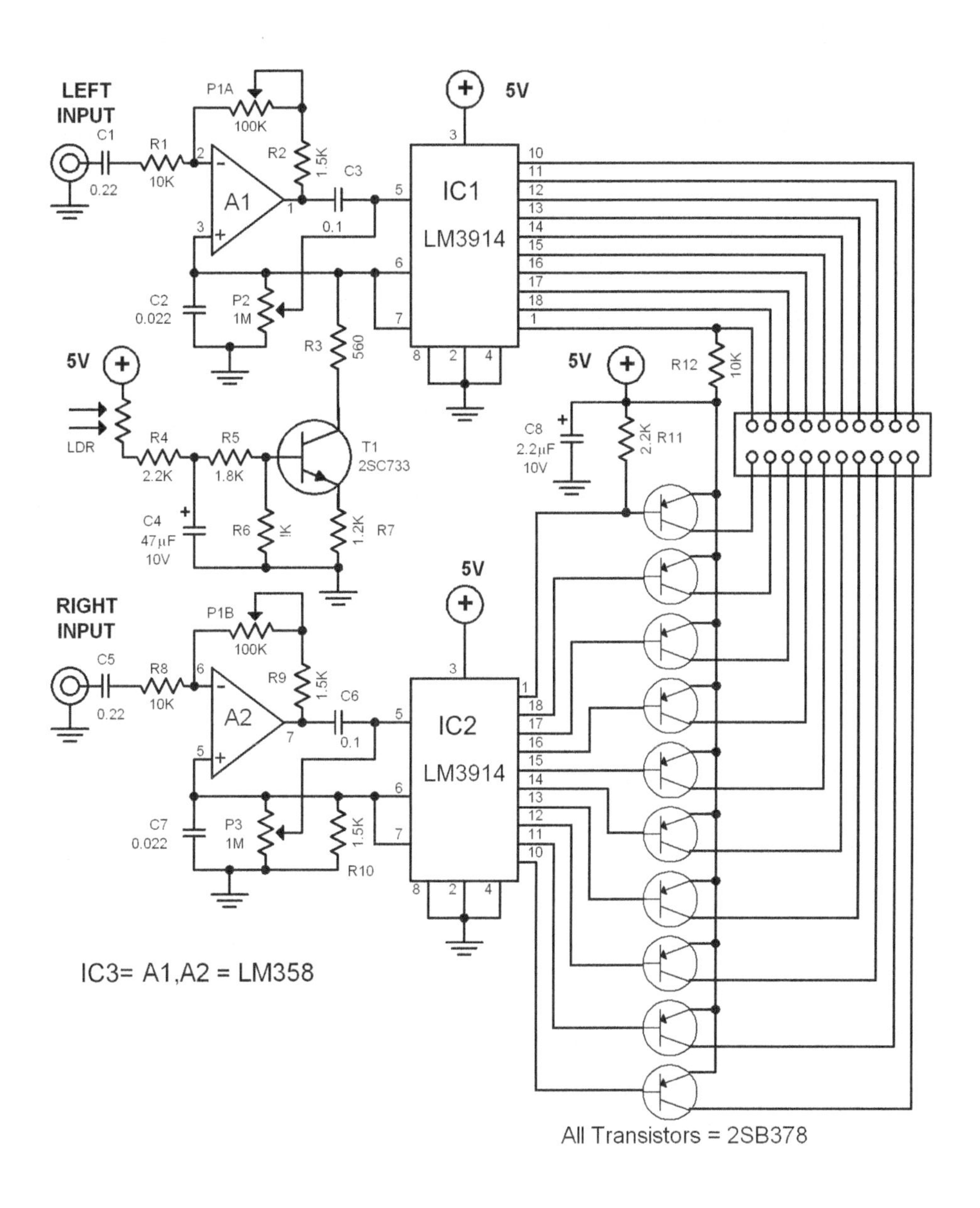

Diagram 23.0 X-Y VU Display

By using an X-Y display matrix, this circuit offers an alternative concept to displaying audio signals which commonly uses a linear two column bar display. The 10 x 10 matrix is controlled through the LM3914 IC from National. The application of the display circuit is varied: it can drive LEDs, LCDs, mini-lamps, and flourescent displays. It can be easily cascaded up to a display resolution of 100 stages.

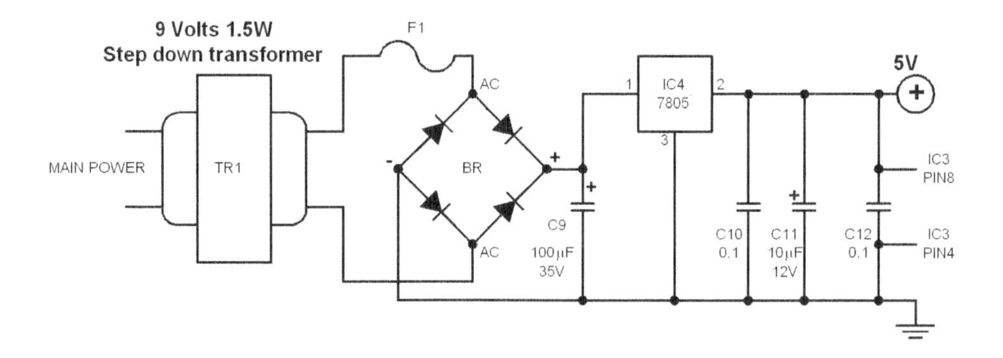

Diagram 23.1 Power supply circuit

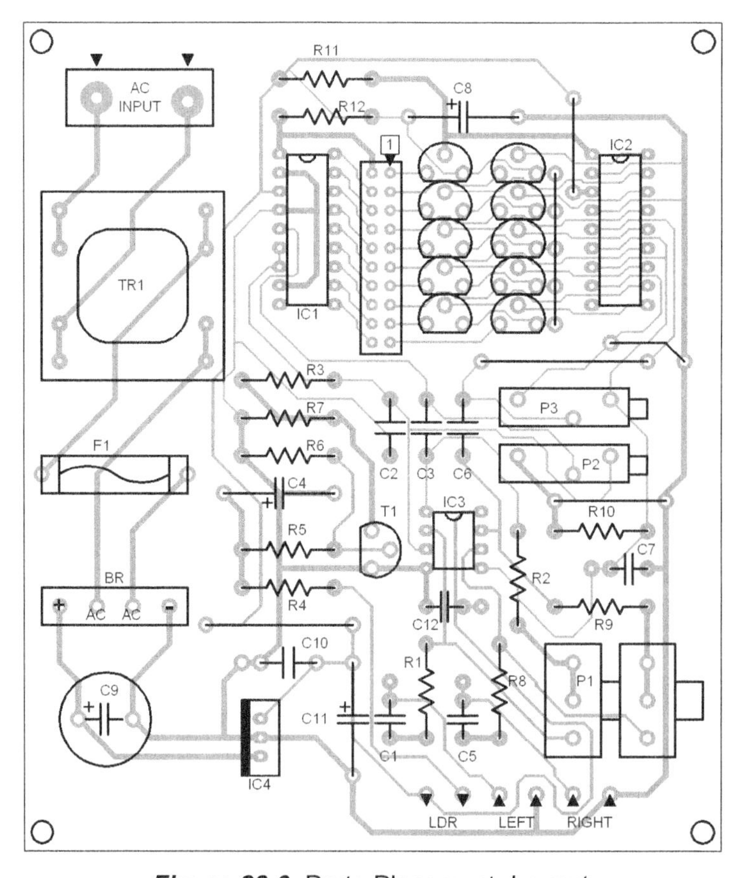

Figure 23.0 Parts Placement Layout

It can be used as a switch between a bar and a dot display. It doesn't need multiplexing. It requires a minimum supply voltage of only 3 volts. Its reference voltage can be adjusted from 1.2 volts to 12 volts. The output current can be set from 2 mA up to 30 mA. It is TTL and CMOS compatible. In the circuit, each LM3914 IC drives the X and Y lines respectively. Since the IC works in dot mode, only one line (X or Y) is driven each time so that only one LED lights up. However, in the transition phase from one LED to another, both LEDs light up for a short time. In X-Y mode, four LEDs light up at the same time.

The brightness of the LED is automatically adjusted to the surrounding light. This is accomplished by the LDR and T1 circuit. Depending on the surrounding light, the LED currents are adjusted between 8 and 25 mA. The audio signals are first processed by the opamps A1 and A2 (for both channels) before they are fed to the LM3914 ICs. Potentiometers P1a and P1b set the maximum amplification and as a result set the upper voltage limit. Potentiometers P2 and P3 set the zero point (it sets the audio signal to a DC voltage).

Calibration: Short the audio inputs so that there is no signal entering the circuit. Adjust P2 and P3 so that the four LEDs in the middle light up. Now connect the input to an audio source. While playing a loud music, adjust P1a and P1b so that the outer LEDs light up when the audio is at its loudest.

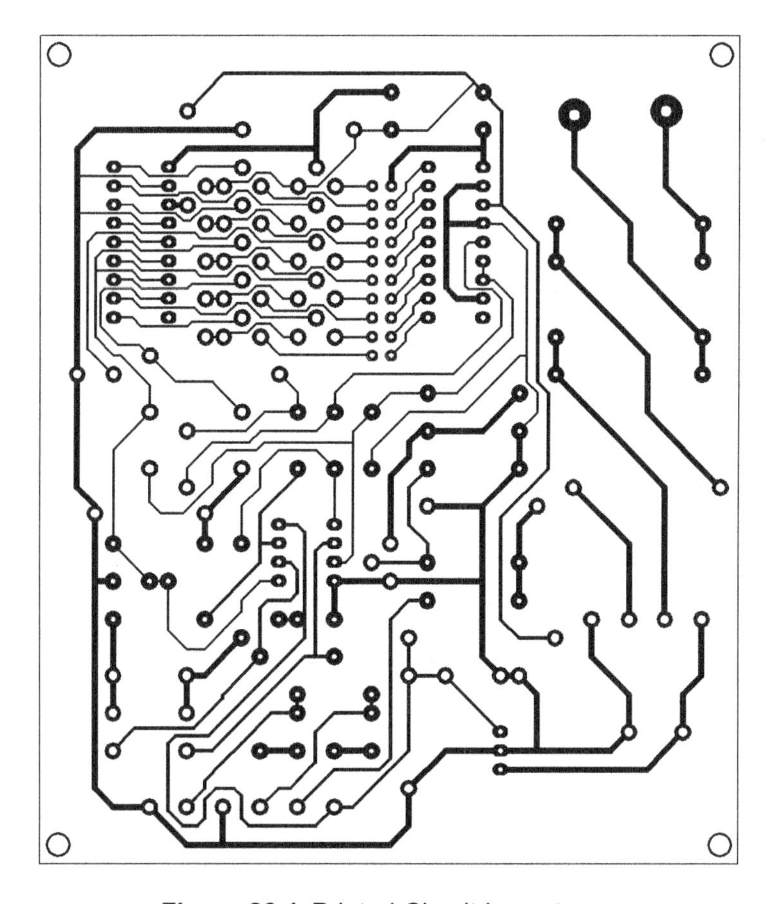

Figure 23.1 Printed Circuit Layout

24 SUBSONIC FILTER

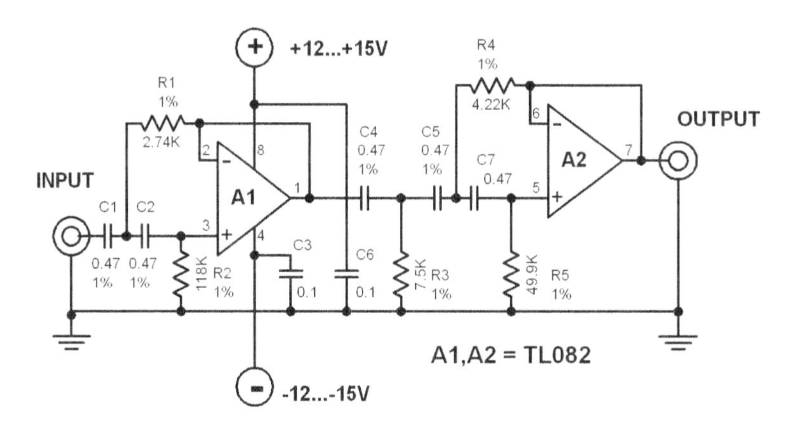

Diagram 24.0 Subsonic Filter

A subsonic filter is mostly used in low frequency technology. Most often, the very low frequencies create distortions in music recording or reproduction. This happens independent of the applied recording medium (e.g. CD, tape, vinyl record, etc). To remove this unwanted low frequencies, subsonic filters like the electronic circuit featured here is used.

The circuit is actually an active Tschebyscheff high pass filter of the fifth order. It has a cutoff frequency of 18 Hz. This filter has a very sharp cutoff character. Frequencies below 10 Hz are attenuated by more than 35 dB. The typical Tschebyscheff resonance of 0.1 dB does not cause any distortion. The phase shift is neglegible.

If the circuit is applied to a stereo unit, build an exact duplicate for the other channel. Any difference could cause noticeable phase shift in the output signal. It is very important that the capacitors C1, C2, C4, C5, and C7 are of exacly the same values in both channels.

The current consumption is around 5 mA. The maximum useful frequency is 3 MHz. In connecting the filter, remove the decoupling capacitor from the front end module to avoid distorting the signal.

25 FULL DUPLEX AUDIO LINE

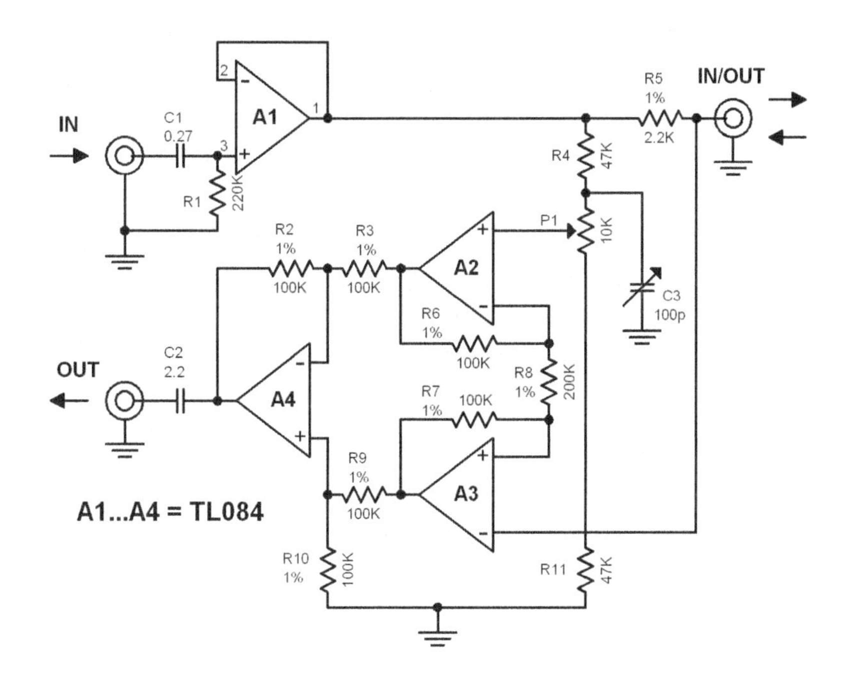

Diagram 25.0 Full Duplex Audio Line

This circuit enables two audio signals in opposing directions to flow simultaneously through a common twisted pair line. And it does that without complicated communications technology. This technique called full duplex is not really new. It is common in telephone technology. However, the telephone uses a carrier signal. This circuit on the other hand does the trick without a carrier signal.

The principle is simple: two transmitters, each one attached at each end of the line, feed the signals. The voltage at the line will be the sum of the two signals: U1 + U2. In the actual circuit however, this is equivalent to only the half of the sum. At each end of the line, the signal is recovered while the other signal gets rejected. This requires that two identical circuits must be constructed.

The opamp A1 works as an impedance converter and at the same time functions as the transmitter. The resistor R5 protects the output of A1 from the signals coming from the transmitter at the other end of the line. The sum of the signals is taken from the output of the circuit and fed to the non-inverting input of the differential amplifier composed of the opamps A2, A3 and A4.

The „right“ signal goes through the voltage divider network made of R4, R11 and P1. It is then taken out from the sum of the signals. This „right“ signal appears at the OUT output of the circuit.

The resistors must be low tolerance type to achieve the highest same phase suppresion possible. This suppression is about 80 dB at 1 kHz and 60 dB at 20 kHz. When using long cables, the suppression can be improved by adjusting C3.

Calibration: Connect the (IN) input to a signal generator and feed a sinus wave of 5 kHz with 1 Veff to the circuit. Attach the twisted pair line (cable) to the (IN/OUT) point of the circuit and short the input of the second circuit at the opposite end of the cable. Raise the signal frequency to 10 kHz and adjust C3 to get the best results.

26 ANALOG SWITCH

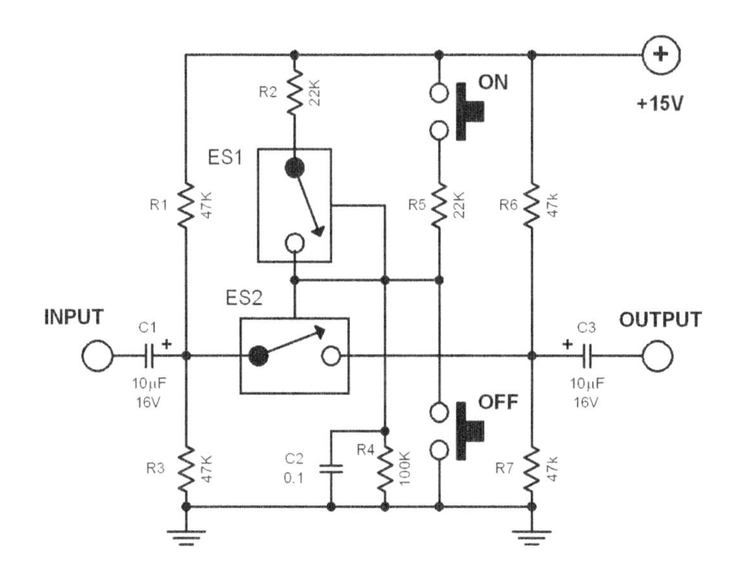

Diagram 26.0 Analog Switch

This circuit is designed to switch on or off an analog line. It consists of two analog switches in IC form that are controlled by two pushbutton switches. In standby mode the voltage at the control pins of the analog switches is low. When the ON button is activated, this voltage increases exceeding the threshold of ES1 and the analog switch closes. Even if the ON button is released, the voltage at the control pin of ES1 remains high and the analog switch remains closed.

This happens because the output of ES1 is connected to its control pin. It therefore latches in the closed position. When the OFF switch is activated, the control pins are connected to the ground and the ES1 opens. This resets the circuit into the standby mode.

In the ON mode, the ES1 controls the ES2 analog switch. The input line is connected or disconnected to the output line through ES2.

27 BATTERY LINE BREAKER

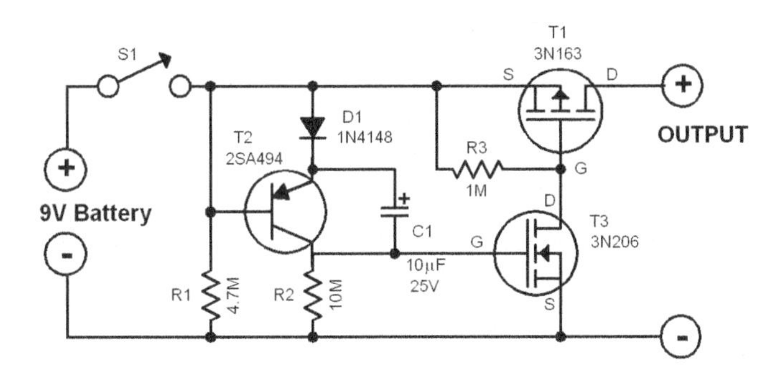

Diagram 27.0 Battery Line Breaker

This circuit is specifically designed to automatically turn off any battery operated device when its user has forgotten to switch it off within a preset period of time. This helps enormously in preserving the battery's life.

How does the electronic circuit work? When the device connected to this circuit is turned on (S1), the capacitor C1 gets a voltage of +9 volts through the diode D1. Since C1 is discharged at the beginning, this voltage goes through to the gate of T3 making both T2 and T3 conduct. The connected device is now turned on as in a normal case. Capacitor C1 then slowly charges via resistor R2. After two or three minutes, the gate voltage of T3 is much reduced to a level that T3 turns off. At this time, T2 turns off the power output to the connected device.

The transistor T1 functions as a quick discharger of capacitor C1 when the user manually switches off the device. It works this way: When the switch S1 is opened (turned off), the base of T1 receives a voltage from C1 via R1 and R2. T1 conducts thereby discharging the capacitor C1. The circuit is again ready for the next switching on. Without this add-on circuit, the user would wait for another minute or two before he can use the device.

Using the circuit is simple. To build it in the device, the positive line of the power (battery) to the device must be cut and the circuit (S1 and output) connected in series to this positive line. The ground line of the circuit must be connected to the minus line of the device.

This circuit can handle current consumptions up to 150 milliamperes. Originally, it was used to control a multimeter device.

28 1-IC VIDEO MODULATOR

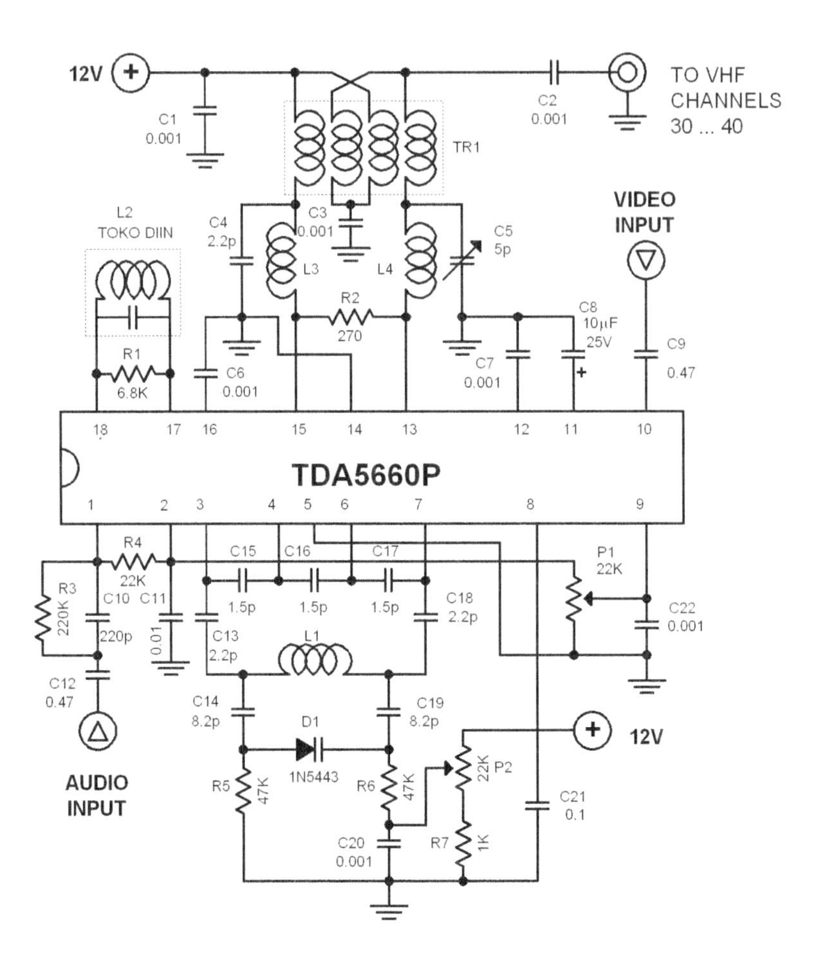

Diagram 28.0 1-IC Video Modulator

The average TV set receives its signal via the standard antenna input. For electronic hobbyists who use a TV set to display signals or old computers that has only a video output, a video modulator is needed. The picture quality is very dependent on the quality of the video modulation. The signals are in HF range and amplitude modulated. A cheap modulator circuit using a single transistor as the oscillator and modulator at the same time does not deliver the needed modulation quality. Better circuits are those that have a separate modulator stage using extra components. A lot better are the circuits with modulator stages that are more advanced in design.

The TDA5660 IC integrates all the necessary functions of an excellent modulator in one circuit. It is specially well suited for hobbyists since it only needs passive components to make it work. The IC can do almost anything a modulator has to do. Its oscillator can generate a signal from 48 MHz up to 860 MHz. The modulator signal can either be positive or negative. The modulation depth is adjustable and the peak and sychronization levels are regulated automatically. The audio modulation can be done either in FM or AM. The frequency drift of the IC is neglegible and the picture and audio quality is excellent.

The featured circuit is designed to the most common configuration. It works with an FM modulation, negative video modulation, and can be set from channel 30 up to channel 40 in the UHF band via the potentiometer P2. The audio input sensitivity is around 450mV. The video signal at the input is kept regulated between 0.7Vss and 1.4Vss. Its current consumption is around 35mA with a 12 volts power supply.

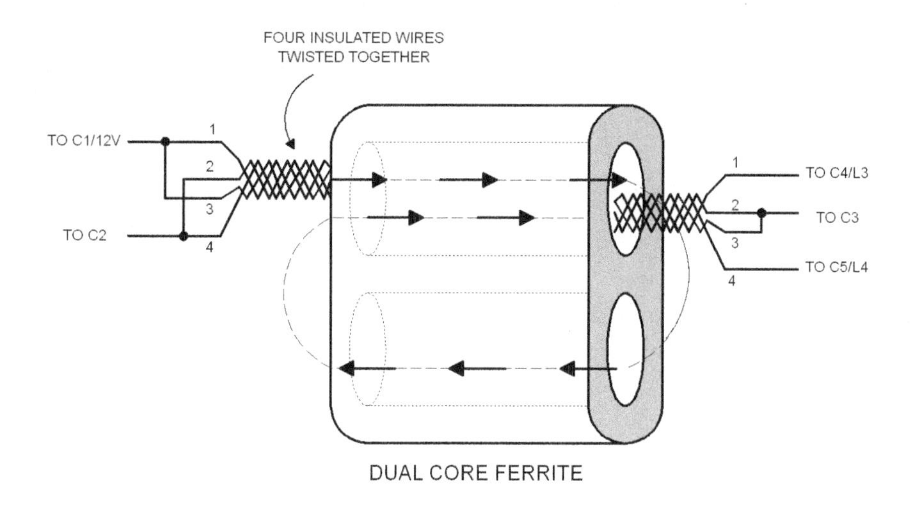

Figure 28.0 Dual core ferrite transformer wiring layout

The preemphasis is done through the R3,R4 and C10. The audio carrier frequency is set with the 5.5 MHz filter (TOKO filter). In case the TOKO is not available, it can be replaced with a 20µH coil and a 50 pF trimmer capacitor. If you wish a 75 ohm input impedance for the video input, connect a 75 ohm resistor between the input and the ground line.

The coil L1 sets the transmitter frequency. It is made of 2 windings of coil wire (around 0.5mm diameter) with a coil diameter of 3mm. A ratcheted potentiometer is recommended for P2. The potentiometer P1 is adjusted to a point of the best audio quality. It is important to use NPO types for the capacitors C13 up to C19.

The complicated part of the project is the symmetical transformer. As shown in Figure 28.0, it is wound in a double core ferrite. The 1,2,3 and 4 wires of the coil must be twisted together before they are inserted in the ferrite core. The L3 and L4 coils are made of 0.3mm diameter copper coil wires. L3 is made of 5 windings and L4 is made of 3 windings, each with a 3mm coil diameter.

29 SYNC SIGNAL SEPARATOR

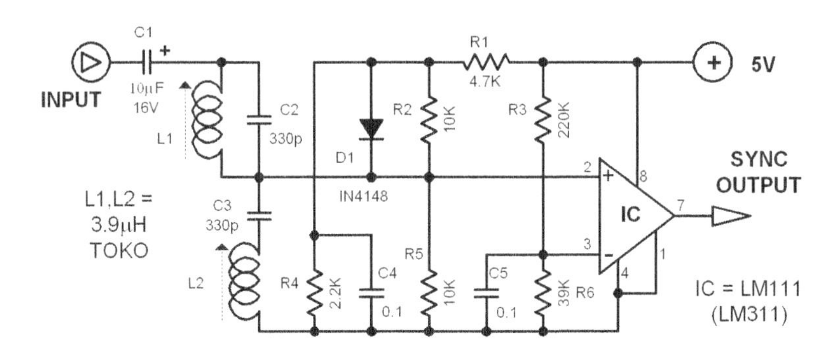

Diagram 29.0 Sync Signal Separator

The composite signal of the TV video is composed of the following: luminance which carries the picture information, blanking pulse, sync signal (composed of horizontal and vertical sync pulses), and the chroma for colored video. The sync signals are important to synchronize the scanning of the picture on the transmitter side with the reproduction of the picture on the receiver side. In a receiver unit, a simple circuit is enough to separate the video signal (luminance plus blanking pulse) from the sync signals. However, if one needs the sync signal to control digital circuits or devices, a more complex circuitry is needed. For example, it is necessary to set the separation level exactly to avoid mistakenly sampling the blanking signal as the sync signal. The blanking pulses are much wider than the sync pulses and this can lead to errors in digital circuits. Another possible source of trouble is the separation glitches that are caused by color or the audio carrier. The color carrier is never fully suppressed and the audio is always present in the signal.

The featured circuit is the result of all the considerations outlined above. The unwanted signals are filtered out using passive components. The filter circuitry at the input filters out the color carrier. The L1 and L2 coils can be replaced with chokes if desired. In that case, replace C2 and C3 with two capacitors made of 270pF and 60pF trimmer connected in parallel. The audio carrier is not filtered out in this circuit. If desired, a 5.5 MHz filter can be added to filter out the audio carrier.

The filtered signal is „clamped“ to about 0.7 volts. This is the DC voltage level where the sync signal rides. This signal is fed to the comparator IC which is the actual filter of the circuit. The minus input of the comparator is set at about 0.75 volts. When the signal at the plus input is below this level, a minus (negative) signal is sent to the output. This sync signal separator circuit delivers a sync signal with a negative polarity. If a positive polarity is desired, just switch the inputs of the comparator.

30 IC TREMOLO

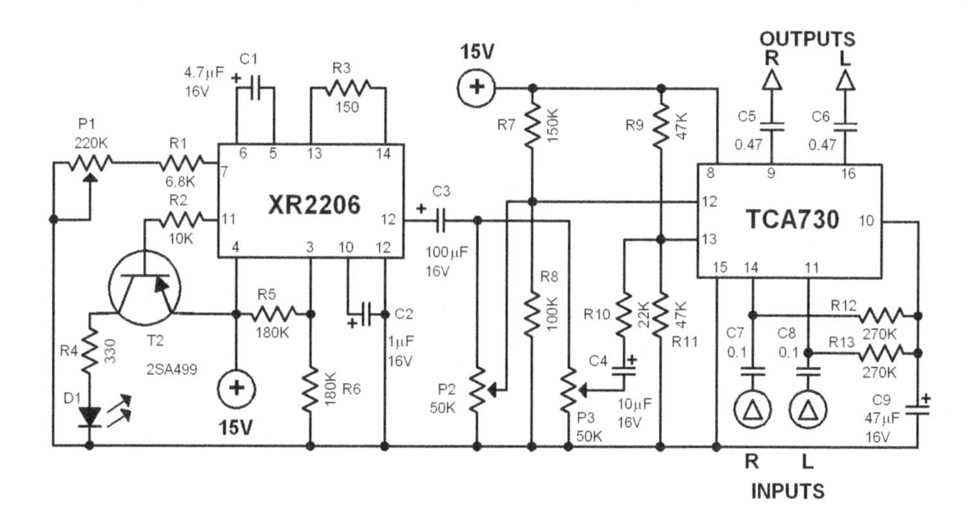

Diagram 30.0 IC Tremolo

Many of the well known tremolo effect circuits have the following disadvantages: the distortions are quite high; the modulation hub and modulation frequency have a narrow bandwidth. The featured circuit allows a modulation hub from 0...100 percent and it is relatively free from distortions. The circuit is useful for stereo channels and it also has the ability to simulate the Lesley effect aka rotating loudspeaker effect.

The circuit uses the TCA730 IC which is designed as an electronic balance and volume regulator with frequency correction. In principle, balance and volume settings are done with a linear potentiometer for both channels. If this potentiometer is replaced with an AC voltage source, a periodic modulation of the input signal can be achieved. This AC voltage source comes from the function generator IC (XR2206). This IC generates square, triangle, and sinewave signals. For this project however only the sinewave is used to create a soft modulation.

The modulation voltage can be varied with P1 from 1 Hz up to 25 Hz. Resistor R3 sets the operational level of the sinewave generator. Resistors R5 and R6 set the DC voltage and the sinewave amplitude at the output. Capacitor C2 is a ripple filter. The squarewave output of the XR2206 drives T2 and a LED to optically display the frequency. The modulating voltage reaches pin 13 of TCA730 via P3 and R10. This input functions as the volume control or in this case the volume modulation. The degree of the balance modulation (Lesley effect) can be varied with P2. A regulated power supply using a 7815 IC is recommended. Do not use a non-stabilized power supply since the current variations would influence the modulation negatively. Attach the 7815 IC to a good size heatsink (about 10-cm^2).

31 AUTOMATIC VOLUME CONTROL

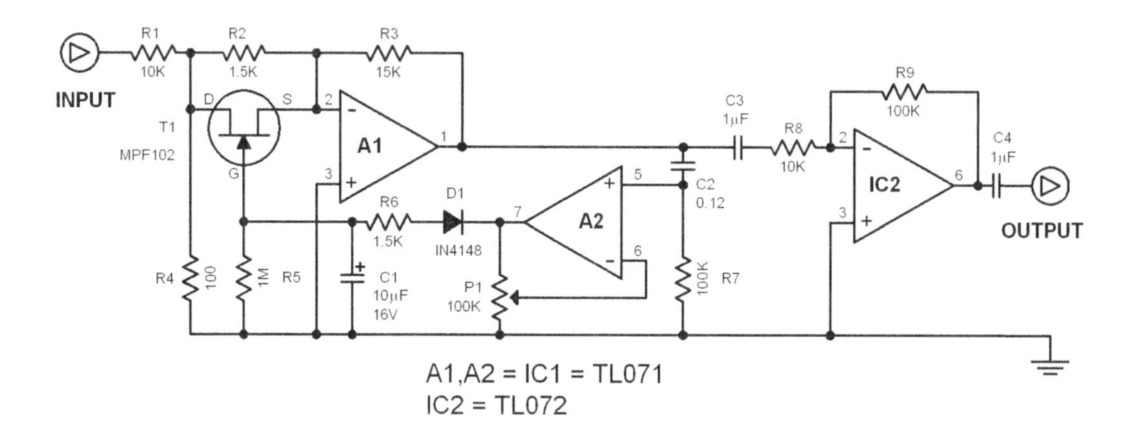

Diagram 31.0 Automatic Volume Control

The function of this electronic circuit is to amplify weak signals without distorting its dynamic compression. The amplitude differences in the signal are levelled off and the disturbing effect vanish. With this technique, overcompensation in the volume is avoided. The circuit is used to automatically control the volume levels of cassette recorders, audio tape recorders, amplifiers, or radio devices.

The circuit works this way: The FET (T1) is used as a variable resistor. The resistance between the drain and source of T1 can be between 150 ohms and infinite. It is parallel to R2 and together with R3 controls the gain of opamp A1 (around 20 dB). The following opamp A2 is wired as an amplifier with P1 as its gain control. The negative part of the output signal coming from A2 is rectified and fed to the gate of T1. Small variations in the signal amplitude does not influence the amplification since the FET has a short delay caused by R6. The opposite effect is also slow because of the discharge time of the C1. Both effects result to a smooth regulation of the signal amplitude.

The signal voltage at the gate of T1 must be as low as possible to influence the drain-source resistance. To achieve this, the voltage divider R1,R4 is set at the input line to attenuate the signal by 40 dB. This technique enables signals up to 1 Veff to be processed without a problem and with distortions levels below 0.5%. The signal-noise ratio is over 70dB by an input voltage of 1 Veff. The losses at the attenuator is compensated by the amplification through A1 and IC2.

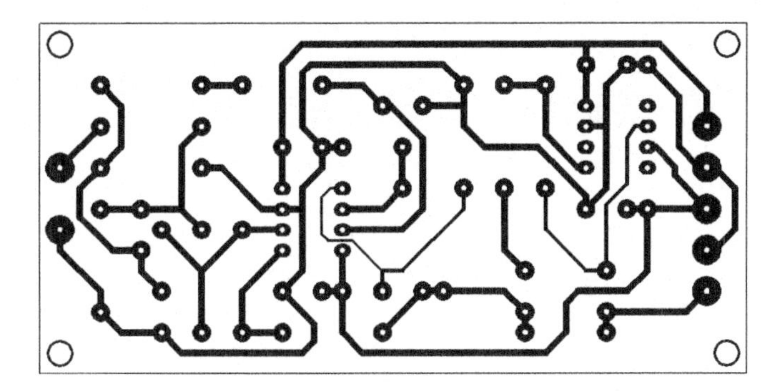

Figure 31.0 Printed Circuit Layout

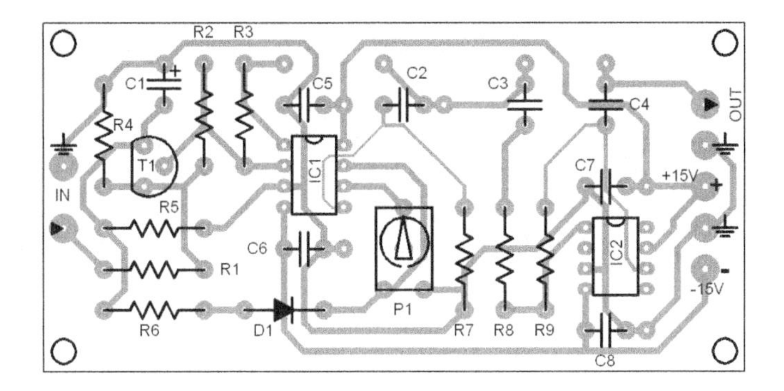

Figure 31.1 Parts Placement Layout

The high pass filter made of C2,R7 prevents the bass signals from influencing the regulation. With proper dimensioning of these components, the cutoff frequency of this filter can be adapted to individual needs. Signals below the threshold set by P1 are amplified by around 18 dB.

The circuit needs a symmetrical supply of +/- 15 volts and consumes around 7 mA.

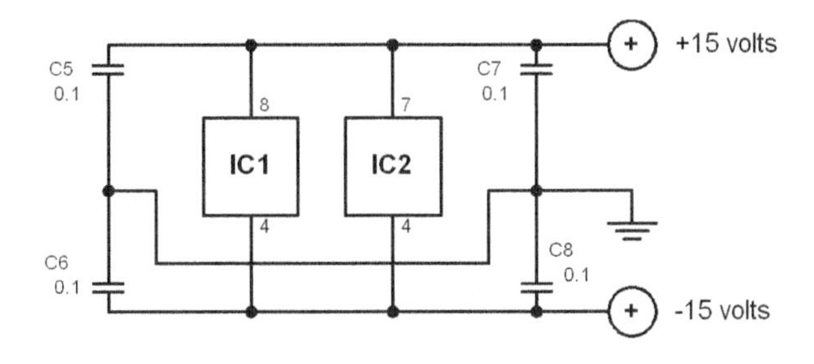

Diagram 31.1 Decoupling capacitors at the ICs' power supply lines.

32 AUDIO TESTER

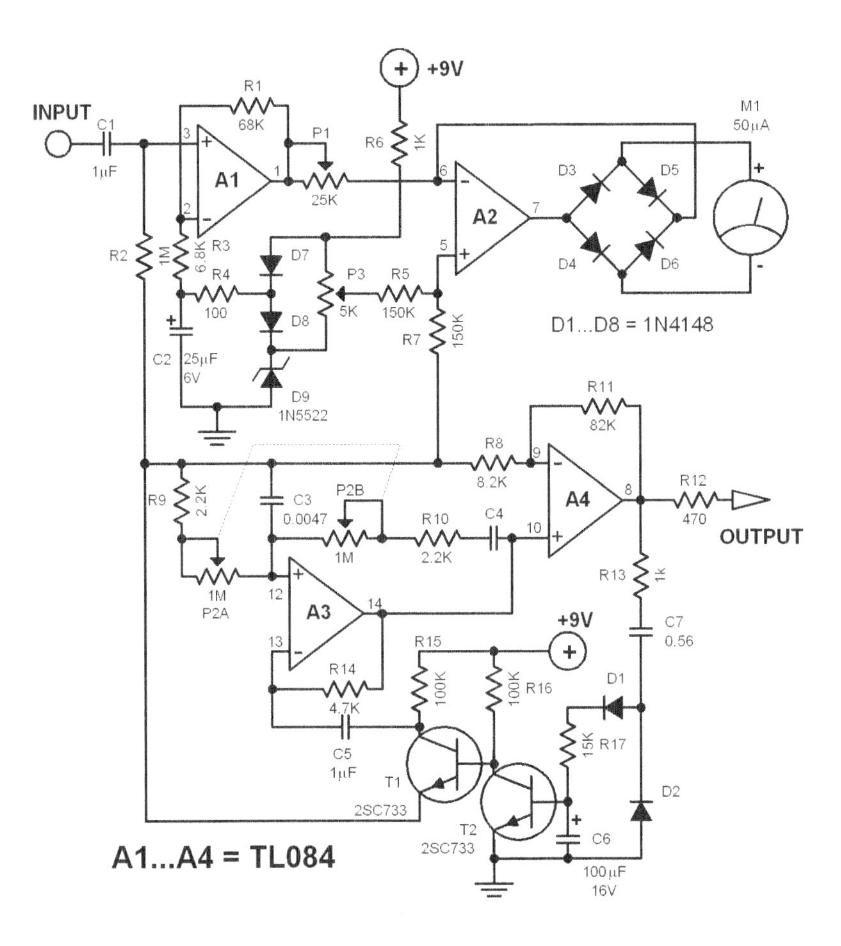

Diagram 32.0 Audio Tester

This circuit is a millivoltmeter plus a signal injector in one package. A simple millivoltmeter combined with a sinewave generator is an ideal device to test audio modules. As shown in the diagram, the opamps A1 and A2 make the millivoltmeter. The lower opamps A3 and A4 make the sinewave generator. The frequency bandwidth is from 150 Hz to 20 kHz.

Since this electronic circuit is powered by a 9 volt battery, the supply line voltage for opamp operation must be divided by two. This is accomplished by the zener diode D9. Resistor R6 is the load resistance for D9. The reference voltage is taken from the D8/D7 junction through the R4/C2 line. This reference voltage is about 5.3 volts. The constant voltage through the two diodes is tapped by P3 and used as variable offset bias for A2. This offset is used to calibrate the millivoltmeter to zero.

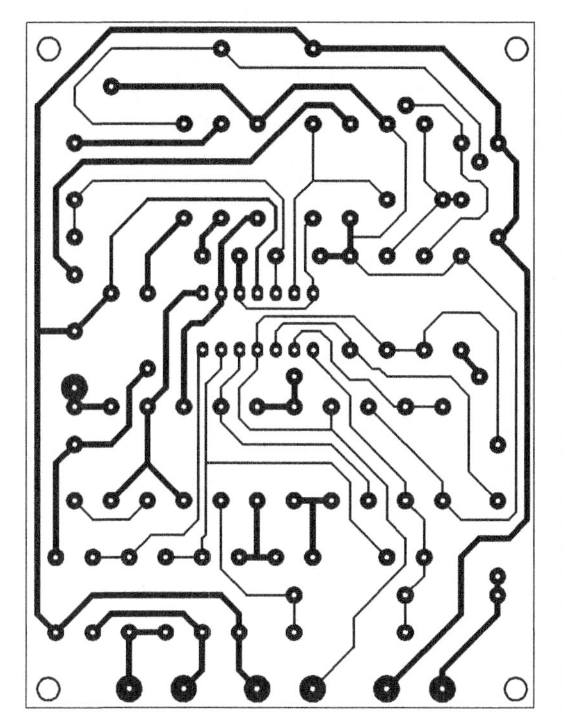

Figure 32.0 Printed Circuit Layout

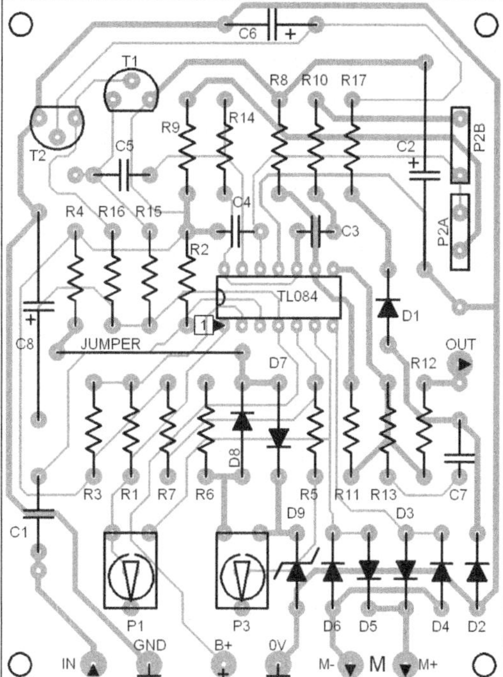

Figure 32.1 Parts Placement Layout

Let us check the millivoltmeter. The input signal enters the plus pin of A1 via the high pass filter created by C1/R2. The input impedance at this point is 1 megaohm. The maximum allowable input level is 50mVeff. To raise this level, one can either add a voltage divider circuit at the input or reduce the gain of A1 by subsituting R1 with a lower value (e.g. for R1 = 6.8 Kohm, the gain is 2 and the input sensitivity is 275mVeff).

The full deflection of the meter M1 is set with P1. The opamp A2 combines with the diodes D3,D4,D5,D6 to create a fullwave rectifier.

The signal generator is basically a wienbridge oscillator with P2 and the capacitors C3 and C4. To stabilize the sinewave, the output signal is tapped from the buffer amplifier A4, rectified via D1,D2,C6,C7. Eventually it is fed to the minus input of A3 via the buffer transistors T1 and T2. With this technique, a relatively stable amplitude of about 2Vpp is generated.

The meter M1 can be any type from 50 µA up to 1 mA. However, the value of P1 in the diagram is chosen for a 50µA meter. To use other current values, convert the value of P1 linearly. For example, for a 500µA meter use 2.5Kohm for P1.

Calibration: take the reference voltage and reduce it to about 45mV by using voltage dividers. Feed this lower voltage level to the plus input of A1 and turn P1 until the meter displays „45" mV.

33 VIDEO SWITCH

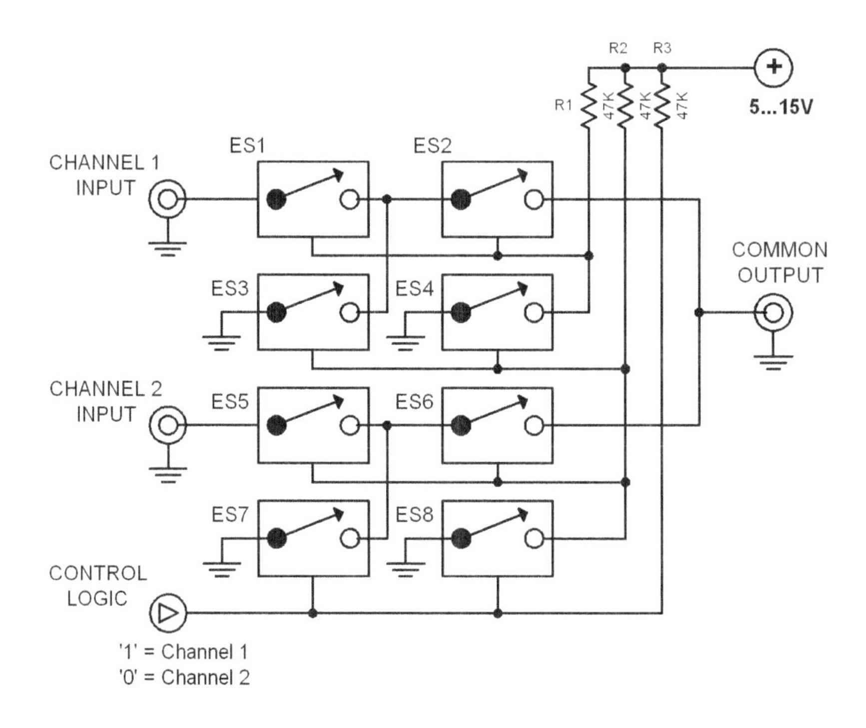

Diagram 33.0 Video Switch

This circuit is designed for switching three different video channels into a single video monitor. This particular design has an advantage over many cheaper models by short circuiting the unused channels. This way, the trailing video signals of a decoupled channel will not appear on the coupled channel.

As you see on the diagram, this is a simple electronic circuit. If channel 1 is switched on, the electronic switches ES1 and ES2 are closed, but ES3 is open. Channel 2 on the other hand is disconnected since ES5 and ES6 are open. Switch E28 prevents any crossover between channels 1 and 2. To increase the decoupling level, each channel uses its own IC.

The output impedance of this electronic circuit is around 75 ohms. The bandwidth is about 8 MHz. The current consumption is dependent on the supply voltage: roughly 1 to 2 mA. A high supply voltage is preferred since the internal resistance of the switches decrease with increasing voltage.

34 3-CHANNEL AUDIO MIXER

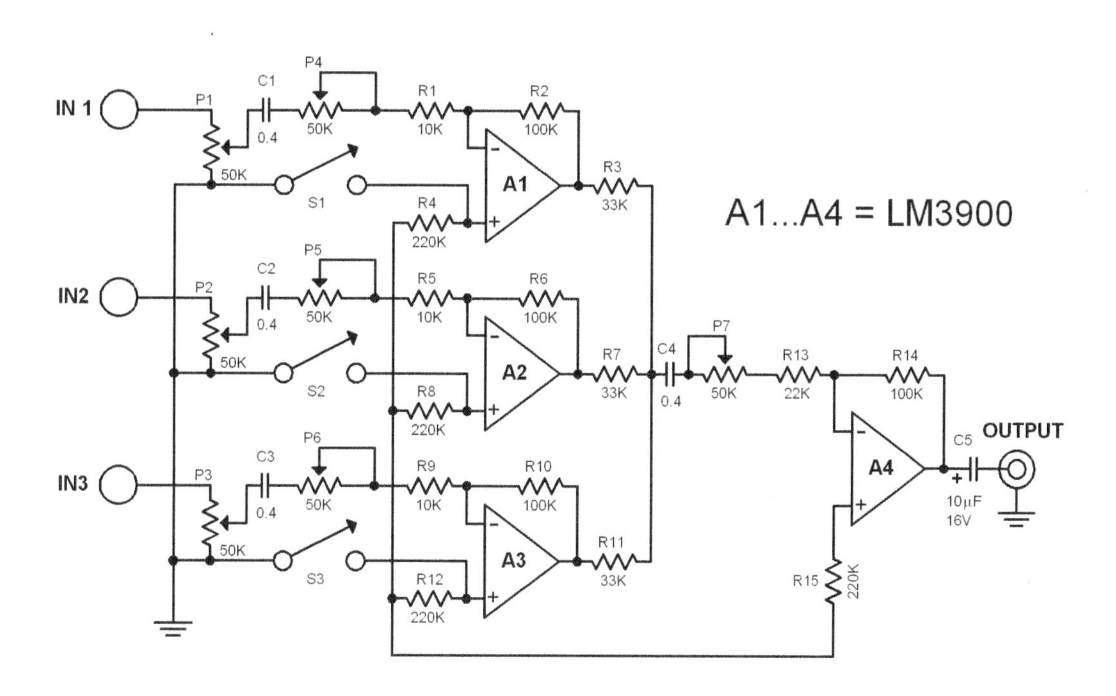

Diagram 34.0 3-Channel Audio Mixer

This mixer circuit uses four Norton amplifiers that are integrated inside the LM3900 IC. The advantage of using this IC is that all four amplifiers need only a single supply. In this circuit, the DC bias is dependent on the feedback since this is a current controlled amplifier. In a Norton amplifier, the output must be set constantly at half of the supply voltage. It allows for a maximum gain without causing distortions. It is called symmetrical limitation at overdrive.

The amplification can be freely chosen. The R2/R1 ratio sets the AC amplification. If R4 is equal to 2R2, then the amplification is null.

Diagram 34.0 shows the complete electronic circuit. The signal level at each input can be varied through P1, P2, and P3. Furthermore, each input can be trimmed to the connected source through the trimmer potentiometers P4, P5, and P6. The resistors at the non-inverting input of the Norton amplifiers set the DC bias and set the output at one half of the the supply voltage. The sum of the signals is then fed to the amplifier A4. The final volume level can be controlled through P7. The switches S1, S2 and S3 either activate or deactivate the input signal.

35 SOUND CONTROLLED LIGHTS

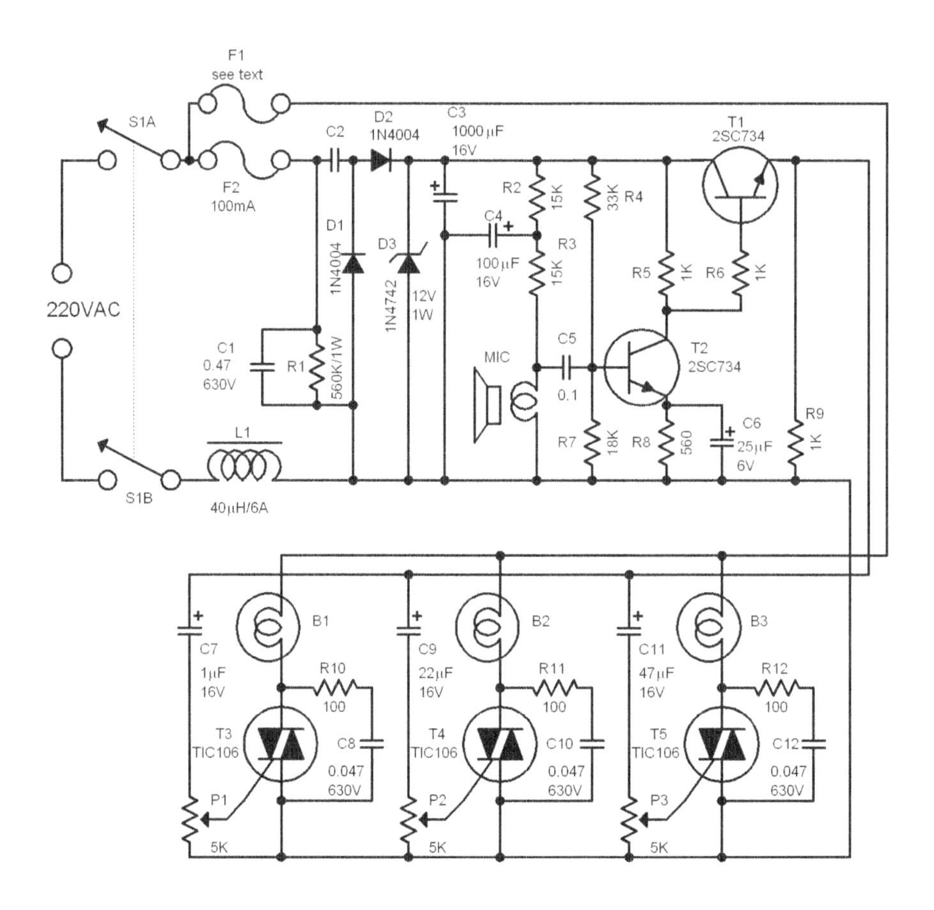

Diagram 35.0 Sound Controlled Lights

This circuit is used to control the brightness of the lights attached to it in sync with the sound that is being captured by its microphone. This electronic circuit is very common in disco houses, bars, parties, and other establishments commonly used for entertainment.

Usually, sound controlled lights are just connected in parallel with the loudspeakers. This configuration has two disadvantages: first, a very powerful amplifer can destroy the lights or worse, a defective light can destroy the amplifier. This problem is avoided by the circuit by not connecting directly to the amplifier. Instead, it picks up the sound with its microphone.

In the featured circuit diagram, the power supply part is on the left of the microphone amplifier and the light controller part is on the right. The capacitors C2 and C3 are the capacitive voltage divider and reduces the power supply level. Diodes D1 and D2 rectify the positive swing of the AC voltage. While D1 conducts the negative swing, the capacitor C3 discharges. The zener diode D3 limits the power supply to 12 volts. Resistor R1 discharges the capacitor C1 when the circuit is unplugged from the main power supply. The network composed of L1 and C1 protects the power line from voltages surges. In the circuit, an electret microphone is being used. Take note that there are two types of electret microphones. The first type has three pins for power, ground, and output. The second type has only two pins. The second type is used for this circuit.

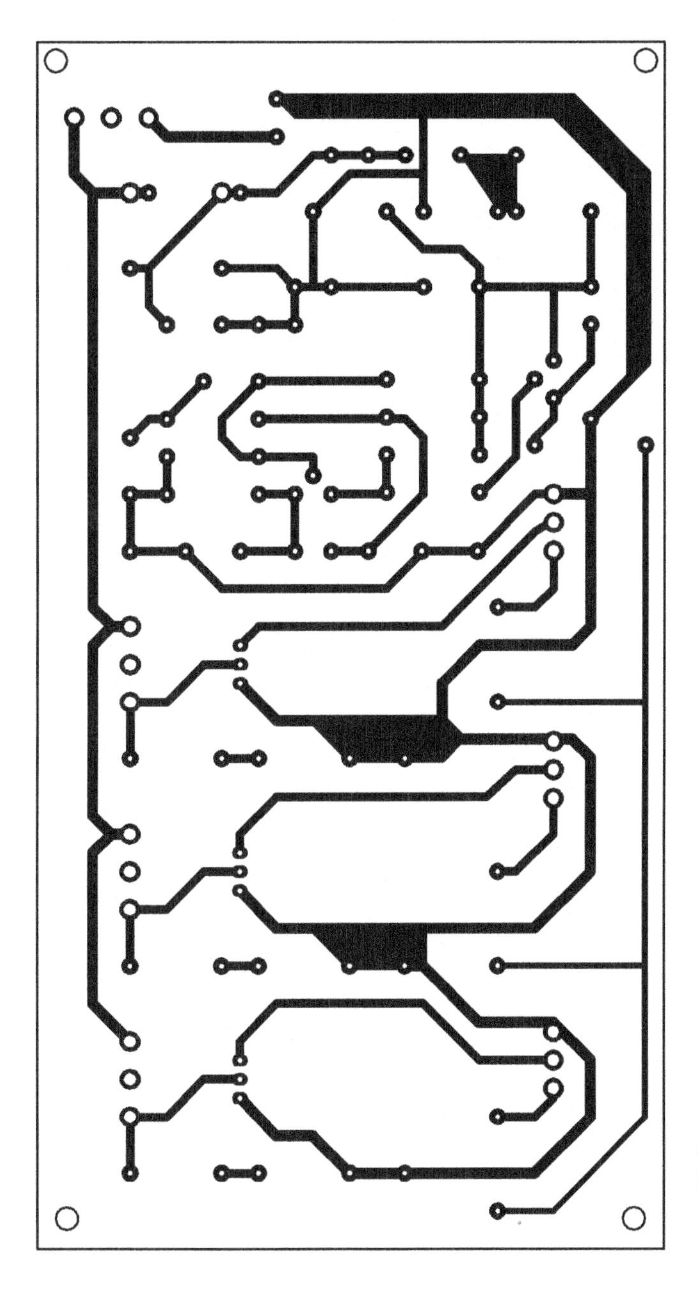

Figure 35.0
Printed Circuit Layout

In building the circuit, take note that you are dealing with strong voltages and currents. The components used must have enough ratings to deal with those. The potentiometers P1, P2 and P3 must have plastic shafts and knobs. Use a shielded cable to connect the microphone to the circuit. All wirings going to the main power supply and lights must be isolated with shrink tubings. The case for the circuit must be made of plastic. Do not let anything metallic protrude out of the box. Even the heatsinks for the triacs are electrified. Beware that there is no isolation between this circuit and the main power lines!

The maximum power output of each output (triac) is 400 watts. Do not exceed that level!

MAINS VOLTAGE IS VERY DANGEROUS. DO NOT WIRE IT UNLESS YOU ARE QUALIFIED. DEATH OR SERIOUS INJURY MAY RESULT.

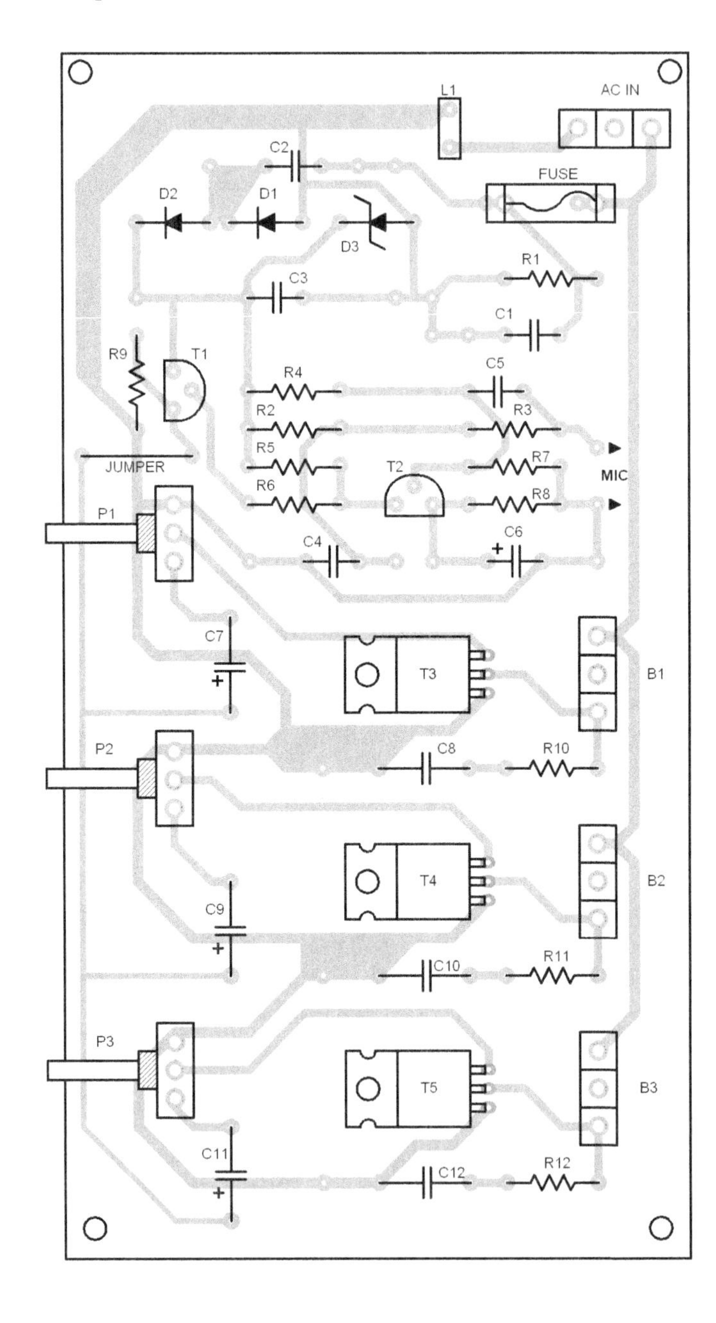

Figure 35.1
Parts Placement Layout

36 AUTO FAN CONTROL

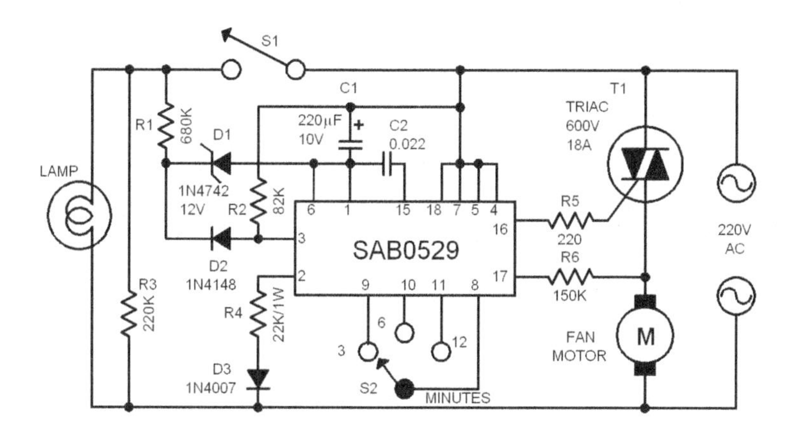

Diagram 36.0 Auto Fan Control

This circuit is a lamp and fan motor control with a delayed switch off. This electronic circuit is commonly used in toilet and bathrooms. When switch S1 is closed, the fan motor and the lamp turns on. When switch S1 is opened (switched off), the lamp goes out but the fan motor will keep running for another 3, 6 or 12 minutes longer (set by switch S2). This delayed switching off of the fan ensures that the unwanted odor is really fanned out of the toilet.

The heart of the circuit is the SAB0529 IC. It is a programmable extended time timer brought to the market by Siemens. This IC can be programmed to time from 1 second up to 31.5 hours. This extended time capacity is achieved by the IC by using the power line frequency as its time base.

The circuit works this way: by closing S1, the IC receives a rectified voltage through diode D3 and resistor R4. The capacitor C1 is a ripple filter. Through this same path, the IC receives its time base signal at pin 2. Pin 3 receives a rising voltage flank and triggers the triac T1 through pin 16. Triac T1 turns on the fan motor. By opening S1 (turning of the switch), the timer starts. The time delay is triggered by a sinking voltage flank. Time delay is selected with the switch S2.

The auto fan control circuit is commonly used to control inductive loads such as the case of fan motor. For this reason, a current synchronization is necessary. This is done with resistor R6. In addition, a capacitor C2 is connected between pins 15 and 1. This capacitor sets the pulse length at the output pin 16. If one decides to use a different triac, one needs to change the value of R5. This resistor sets the level of the trigger current, in this case 5 mA. The sync resistor R6 must be changed also. If the triac power and holding current (Ih) are known, one can get the value of R6 from the table below.

Power	Ih 20mA	25mA	50mA	75mA
100	300K	400K	850K	1M
150	200K	250K	470K	850K
200	125K	175K	400K	650K
250	87K	133K	300K	500K
300	47K	100K	250K	400K
350	33K	75K	200K	350K
400	27K	47K	175K	300K
450	18K	33K	150K	250K
500	10K	18K	133K	220K
550		10K	100K	200K
600		4.7K	87K	175K
650			75K	150K
700			57K	133K
750			47K	110K

37 DX AUDIO FILTER

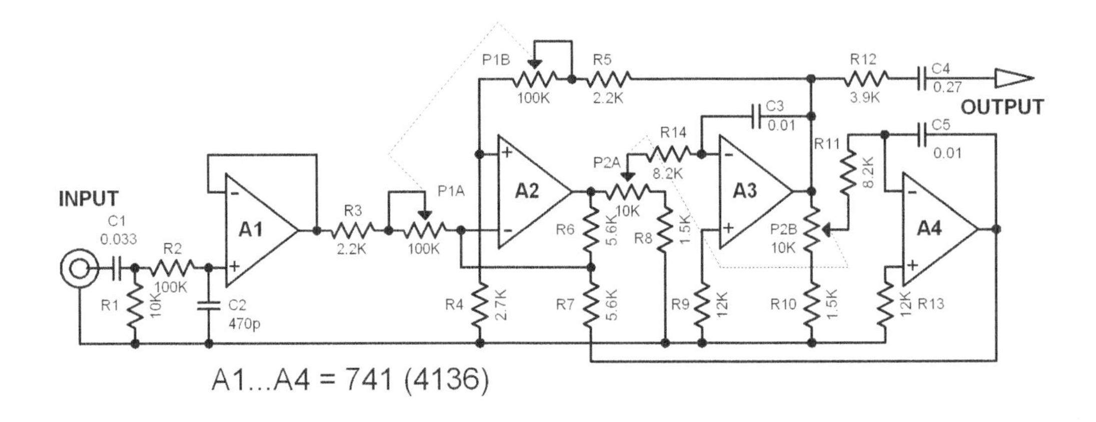

Diagram 37.0 DX Audio Filter

Most universal radio receivers (specially the cheaper ones) have a very wide bandwidth that is not particulary useful for radio amateurs. The better models with narrower bandwidth are almost always priced higher, too high for the reach of an average radio hobbyist.

However, if one wants to use a wide bandwidth receiver to listen to amateur SSB and CW stations with the least amount of interference, this featured electronic circuit can be of great help. This audio band filter cleans out the unwanted noise. It is a state variable type of filter, a bandpass filter with a variable center frequency and bandwidth. If this filter is connected in front of the audio amplifier, one can filter out the unwanted signals from the audio. One requirement for it to work properly is that the filter's center frequency is set exactly to the audio frequency of the wanted signal.

The RC network made of R1,R2,C1 and C2 works as input bandpass filter with a 6 dB cut-off at 500 Hz and 3400 Hz. The opamp A1 functions as a buffer stage between the input filter and the state variable filter. The state variable filter is composed of the opamps A2, A3 and A4. The potentiometer adjusts the bandwidth of the filter. The potentiometer P2 adjusts the center frequency of the filter from about 200 Hz up to 2 kHz. By setting these two potentiometers correctly, it is possible to filter out a certain frequency band from the entire audio spectrum.

The power supply is made of two 15 volts. The circuit consumes very little current.

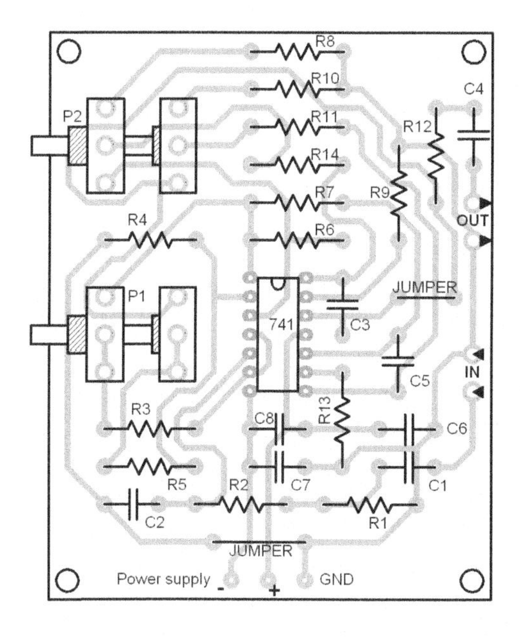

Figure 37.0 Parts Placement Layout

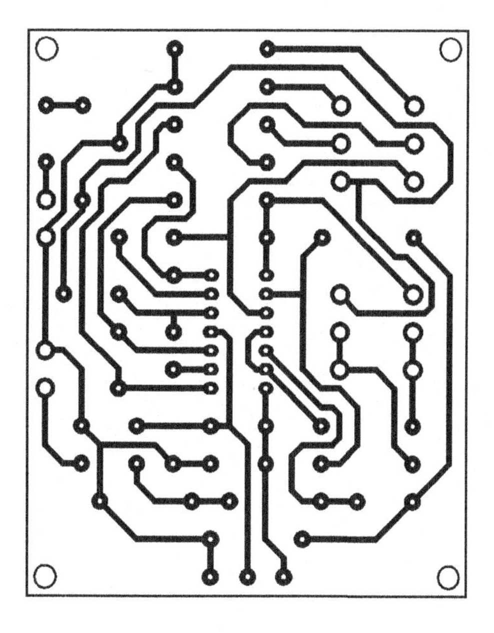

Figure 37.1 Printed Circuit Layout

38 2-Meter Foxhunt Transmitter

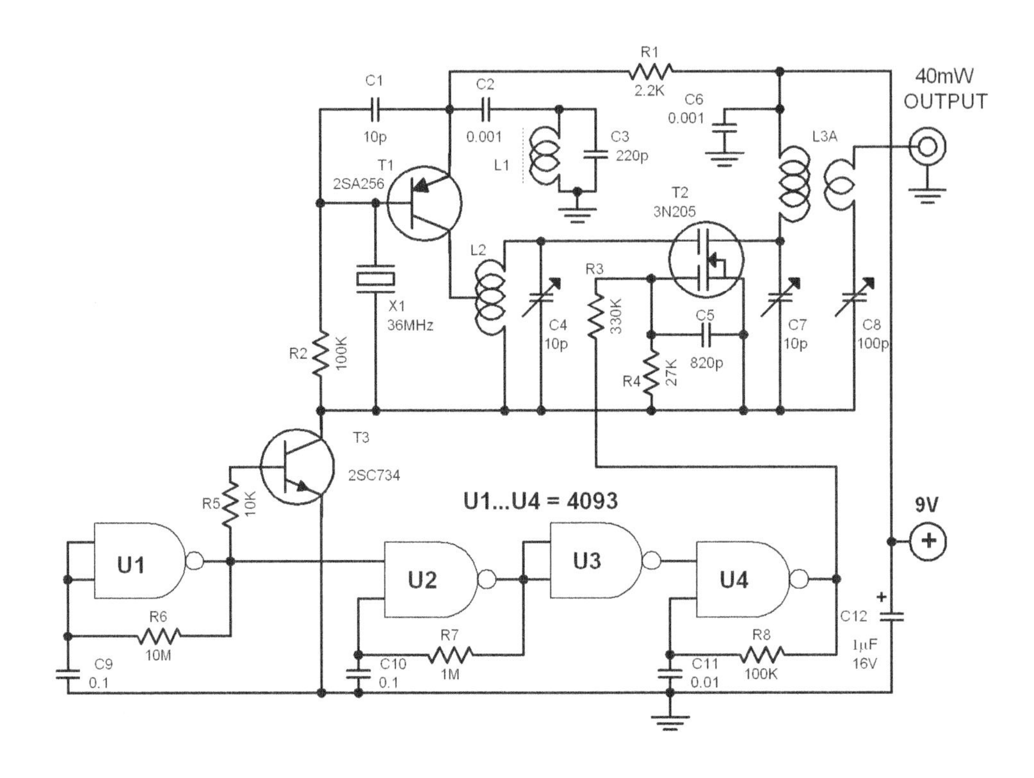

Diagram 38.0 2-Meter Foxhunt Transmitter

This circuit is commonly used in amateur radio competitions where a hidden transmitter is to be „hunted" using mostly homebrewed receivers and antennas. The featured electronic circuit is the transmitter. It radiates a high quality signal without unwanted harmonics. Transistor T1 and the crystal together make the oscillator that generates a 36 MHz (third overtone) signal. The unwanted 12MHz basic frequency of the oscillator is suppressed by the filter circuit made of L1,C3,C2. The L2/C4 circuit is set to the fourth harmonic or 144 MHz. The signal goes to the dual-gate-FET driver stage before finally radiating through the transmitting antenna. The output power is from 10...40 milliwatts.

The radiated signal is also modulated by the gate circuit made of U1,U2,U3,U4. Gate U1 is a low frequency oscillator that generates a signal from 0.1 ... 0.5 Hz. This signal modulates the transmitter through the transistor T3.

If the U1 output is „0“, transistor T3 is off and the transmitter is also off. On the other hand, if the U1 output is „1“, transitor T3 is on and the transmitter is on. During the „1“ period of U1, gate U2 generates a squarewave signal with a frequency from 0.1 ... 1 Hz. Gate U3 works as an inverter only. It determines whether gate U4 generates a 1 kHz signal (U3 high) or not (U3 low). A periodic burst signal is now present at the gate of the FET T2 to modulate the transmitter. One can decide to change the resistor values at the gate oscillators to change the modulation values. This is sometimes necessary to differentiate several transmitter from each other. Otherwise, it is impossible to hunt several transmitters transmitting simultaneously with the same modulation.

Calibration:

Adjust the three trimmer capacitors to produce a maximum signal amplitude at the output.

Coil data:

L2 is made of 5 windings of 0.8mm copper wire, 8mm winding diameter. It is tapped at the first winding from the ground.

L3 is made of 0.8mm copper wire, 8mm winding diameter, 3 windings at the FET side and 2 windings at the antenna side. Adjust the coupling between the two winding sides to get the maximum signal output amplitude.

The circuit can be powered with a 9 volt battery. It consumes around 20 mA only.

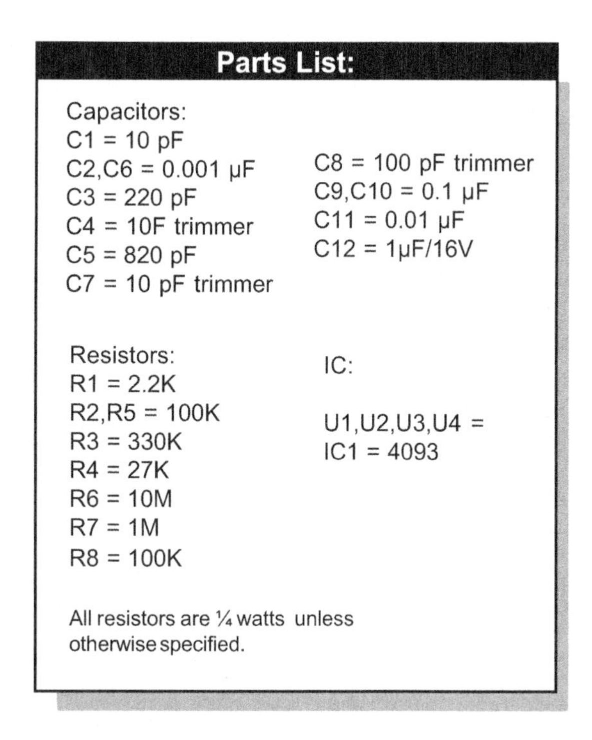

Parts List:

Capacitors:
C1 = 10 pF
C2,C6 = 0.001 μF
C3 = 220 pF
C4 = 10F trimmer
C5 = 820 pF
C7 = 10 pF trimmer
C8 = 100 pF trimmer
C9,C10 = 0.1 μF
C11 = 0.01 μF
C12 = 1μF/16V

Resistors:
R1 = 2.2K
R2,R5 = 100K
R3 = 330K
R4 = 27K
R6 = 10M
R7 = 1M
R8 = 100K

IC:
U1,U2,U3,U4 =
IC1 = 4093

All resistors are ¼ watts unless otherwise specified.

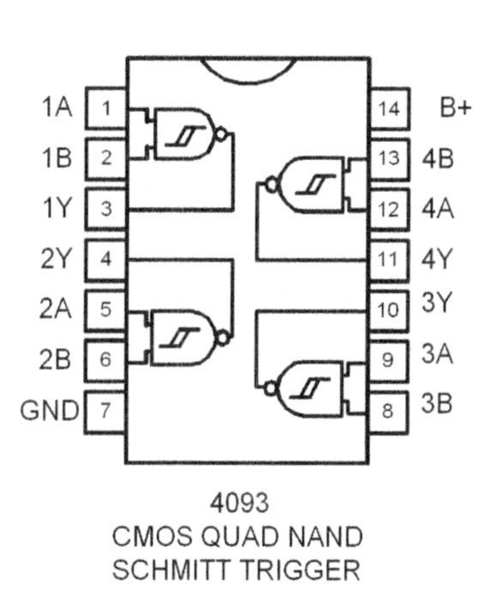

39 UHF ANTENNA PREAMP

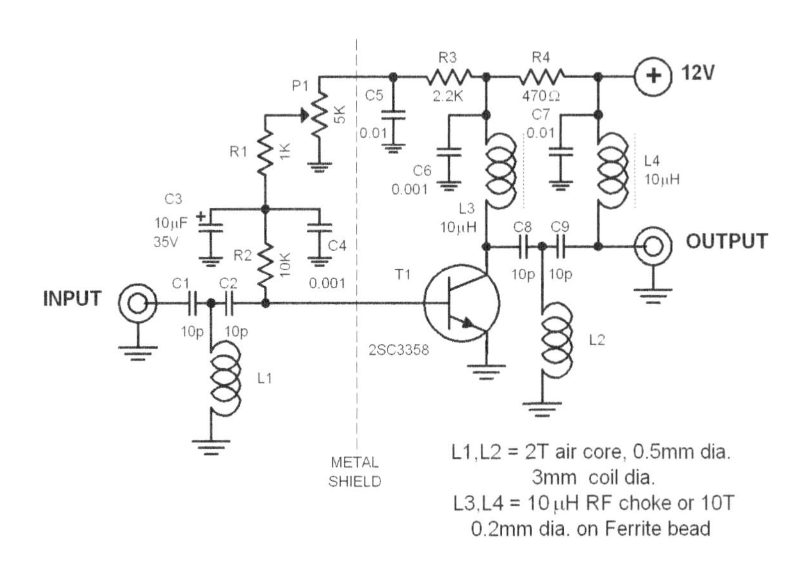

Diagram 39.0 UHF Antenna Preamp

Receiving television signals is sometimes plagued by problems caused by inadequate signal strength. In such cases, using a tv antenna amplifier is necessary. The featured electronic circuit does the job very well. It is a UHF amplifier made with a single low noise bipolar transistor. The amplifier achieves a gain of 10 up to 15 dB within the bandwidth of 400 ... 850 MHz. The recommended pcb design is shown on the following page. The transistor is in the traditional dual-gate-mosfet packaging. It is commonly mounted in a hole in the printed circuit board. This transistor however has two emitter electrodes and both must be soldered to the circuit ground.

As shown in the pcb layout, the pcb is divided in two areas by a thin metallic shield. The transistor is exactly beneath the notch in the middle of the shield. This metal shield is soldered to the circuit ground. The disc capacitors must be mounted in a way that one side of the disc is directly soldered to copper plate. Keep all wirings short as possible. The input and output capacitors are 10pF SMD variant. They build an output filter together with L1 and L2.

Connecting the circuit directly at the antenna is highly recommended. It must be housed in a waterproofed box.

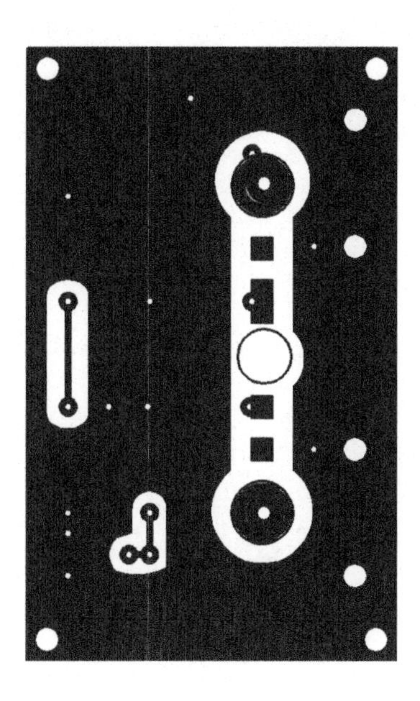

Figure 39.0 Printed Circuit Layout

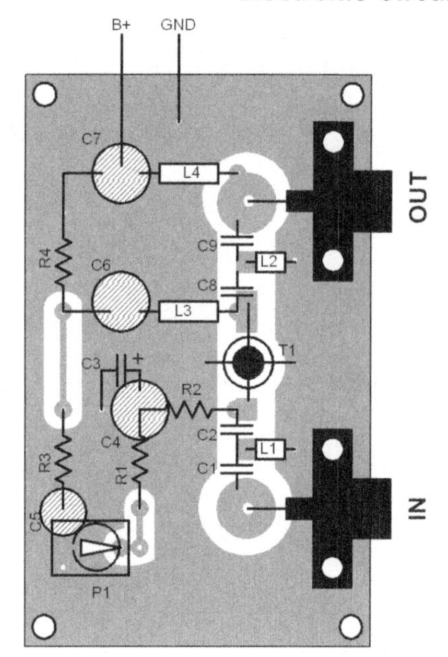

Figure 39.1 Parts Placement Layout

The power supply for the circuit is delivered remotely using the coaxial cable. If remote power supply is chosen, connect the 12 volt power to the coax cable via a 100 µH choke. Connect the power supply ground to the coaxial braid. Connect the coaxial cable to the TV receiver via a 10...47 pF capacitor to prevent the 12 volt power from entering the TV set. See diagram 39.1. If you decide not to use a remote power supply, remove L4 from the amplifier circuit.

Calibration: Adjust P1 to get the best reception.

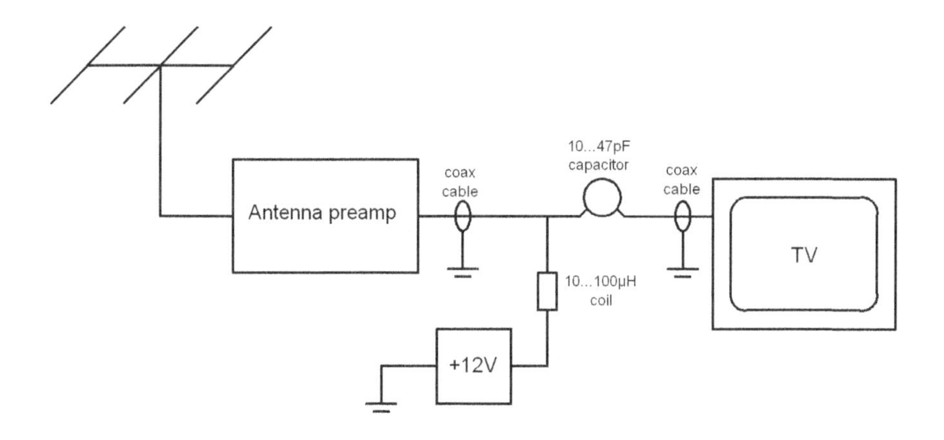

Diagram 39.1 Connecting a remote power supply via coaxial cable

40 FSK FILTER

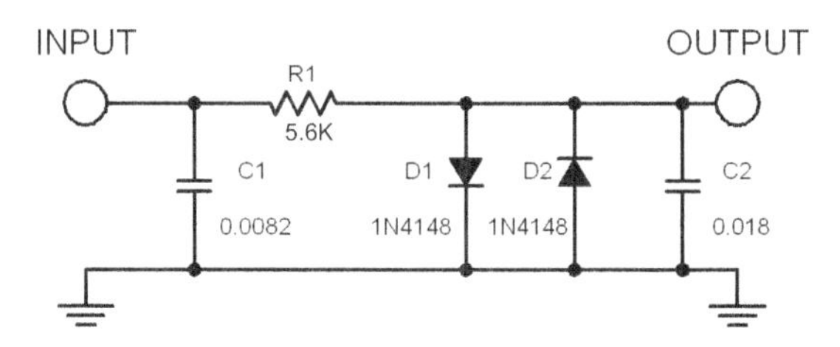

Diagram 40.0 FSK Filter

Frequency-shift keying (FSK) is a form of frequency modulation in which the modulating signal shifts the output frequency between predetermined values. Usually, the instantaneous frequency is shifted between two discrete values termed the mark frequency and the space frequency. Continuous phase forms of FSK exist in which there is no phase discontinuity in the modulated signal. The example shown at right is of such a form. Other names for FSK are *frequency-shift modulation* and *frequency-shift signaling*.

If problems arise in demodulating FSK signals, it is helpful to use this featured filter circuit. The circuit is too simple that it can even be built inside a phone plug. The diagram shows a low pass filter made of R1 and C2. This filter has a cutoff frequency of about 1600 Hz. Since the „0" and „1" signals of the FSK are sent as tones of 1200 Hz and 2400 Hz respectively, the filter removes all unwanted corners from the signal.

Both diodes D1 and D2 ensure that the FSK signal is limited to about +/- 600 mV amplitude. The FSK filter can be attached to a loudspeaker or headphone output.

41 FREQUENCY CONVERTER

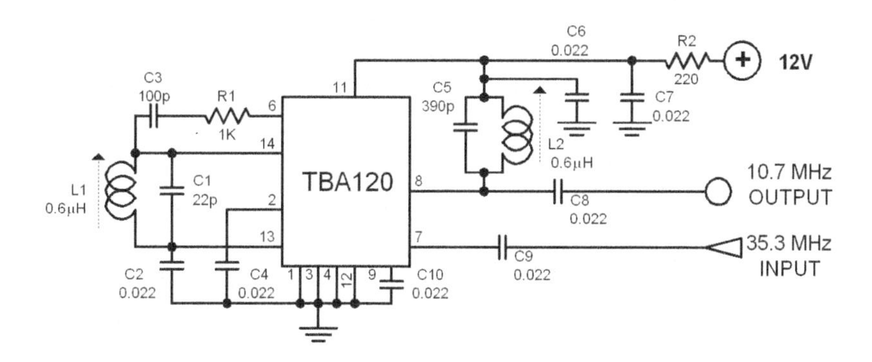

Diagram 41.0 Frequency Converter

This is a 35.3 to 10.7 MHz converter circuit. It converts the 35.3 MHz signal coming from a VHF/UHF tuner down to an FM tuner to decode the TV audio in FM quality. The IC being used (TBA120) is one of the most common components in radio and TV circuits. Although it was designed as an RF amplifier and FM demodulator, it can also be used for many other applications. In fact, it was even found in some audio frequency devices!

The circuit works according to a simple principle: a converter has always a mixer and an oscillator. The TBA120 can be used as the mixer. Its amplifier component can then be used as the oscillator by adding an LC network to it (L1/C1).

Using the values given on the diagram, the circuit oscillates at 46 MHz. This is then mixed with the input signal of 35.3 MHz resulting to a 10.7 MHz output signal. Following this example, one can modify the circuit to convert other frequencies. The components that need to be redimensioned are the L1/C1 oscillator network and the output filter L2/C5. If the desired oscillator frequency is far below 46 MHz, the R1/C3 must also be adapted as well as the L1/C1. Both values must be increased for lower oscillator frequencies but their exact values are not very critical and can be derived experimentally.

The pcb design must be very simple due to the simplicity of the circuit. However, it is highly recommended to provide lots of ground copper plating. Also, keep the copper lines short and narrow. Most specially, keep the copper lines of the decoupling capacitors to ground as short as possible.

42 FIELD STRENGTH METER

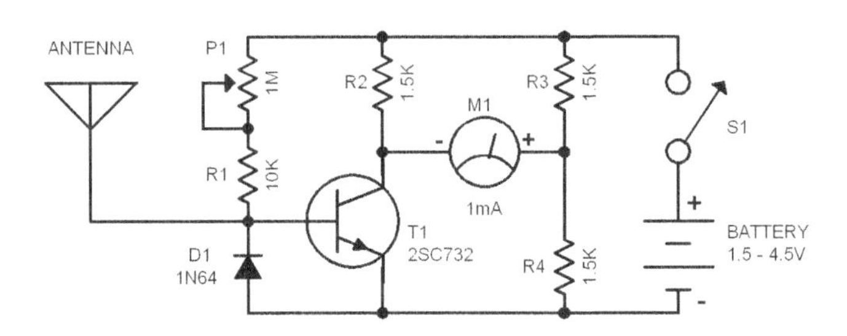

Diagram 42.0 Field Strength Meter

This circuit measures the strength of a radiated signal from a transmitter. It is commonly used by model airplane builders to check whether their transmitters are really transmitting the radio signal. Using this FS meter, one can quickly determine whether the problem lies on the transmitter or the receiver module.

A single transistor T1 is the only active element in this circuit. The transistor is designed as a „controlled resistor“ forming one part of a balanced bridge circuit. The base of T1 is directly connected to the receiving antenna. With increasing radio frequency voltage at the transistor base, the transistor conducts and disturbs the balance of the bridge circuit. With the bridge out of balance, current flows to the meter through R3, the meter and the emitter-collector junction of the transistor.

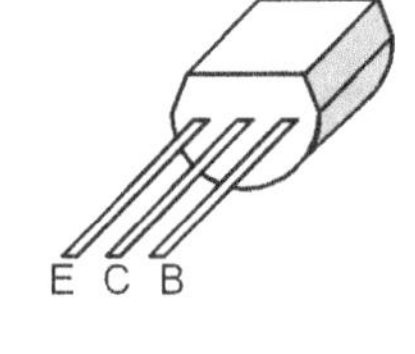

2SC732

How to use: Set the meter to zero by adjusting P1 before you turn on the transmitter.

43 AUTOMATIC VOLUME CONTROL

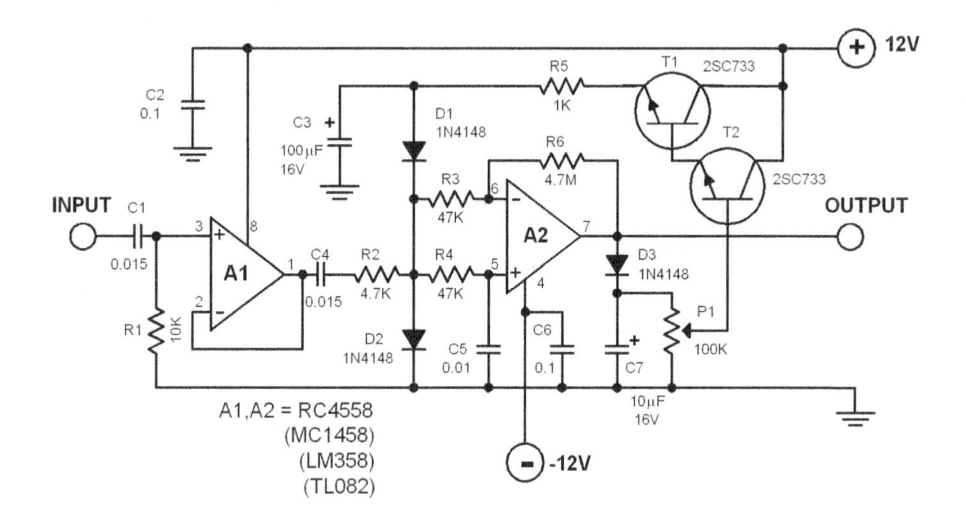

Diagram 43.0 Automatic Volume Control

The featured circuit controls a volume line automatically. It delivers an output voltage of approximately 4 volts peak to peak. This voltage remains relatively constant by input voltages ranging from several hundred millivolts to several volts. This electronic circuit is not highly recommended for hifi applications because its noise factor is way above the accepted level. However, it is very useful in tape recording of computer programs. The constant amplitude is desirable in such application.

The opamp A1 works as an input signal buffer. If we remove the diodes D1 and D2, this opamp will work as an amplifier. The DC bias of A1 is done over R4 and C5. This little trick allows the A2 to limit the DC level at its input to a maximum amplification of 100 times . The offset bias is relatively constant. The amplified signal from A2 is rectified and filtered by D3 and C7. A sample of this rectified signal is fed to the regulator diodes D1 and D2 over the transistors T1 and T2. Trimmer P1 controls this sample signal. The higher this signal is, the higher is the current flowing through the diodes.

These regulator diodes have a non-linear curve. Their resistance decrease with increasing current. The input signal gets grounded more or less through the diodes. To put it in another way: the diodes work as an attenuator with increasing effect by increasing the current through the diodes.

44 ELECTRONIC HEAD OR TAIL

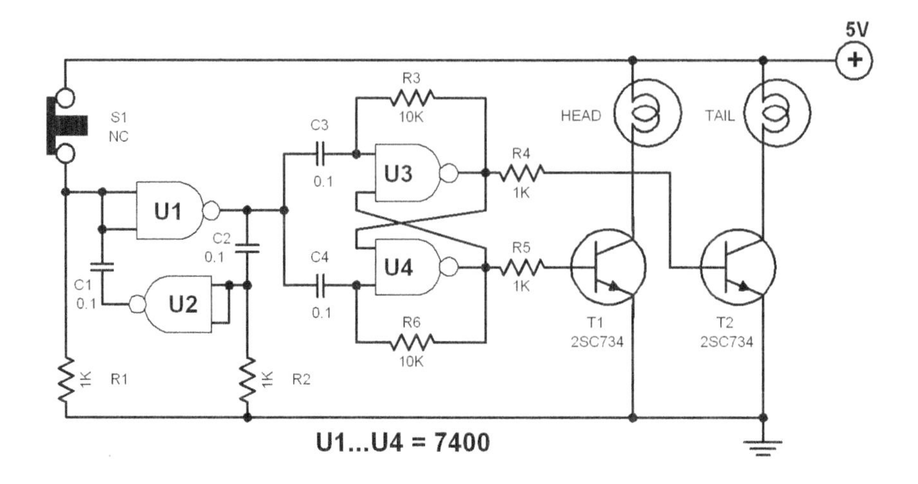

Diagram 44.0 Electronic Head or Tail

The principle used in this electronic head or tail circuit is simple: A multivibrator controls a flip flop. The multivibrator oscillates as long as the button S1 is pressed and the flip flop switches on and off with a frequency of several kilohertz. When the button is released, a +5 volts is set at one of the two gates that make up the multivibrator. The flip flop latches on at one of the two possible states: „head" or „tail".

The state of the flip flop is dependent on the exact time when the button S1 was released. Because the speed of the flip flop's switch overs and the relative inertia of the human reaction, the state at which the flip flop will latch on is random.

The entire circuit is made up of a single 7400 IC. It is important to decouple the circuit and the display bulbs very well to avoid reverse voltages and currents that appear during the switching off of the display bulbs.

The capacitor C5 (not shown in the schematic diagram) is a 250 μF electrolytic capacitor rated at 10 volts.

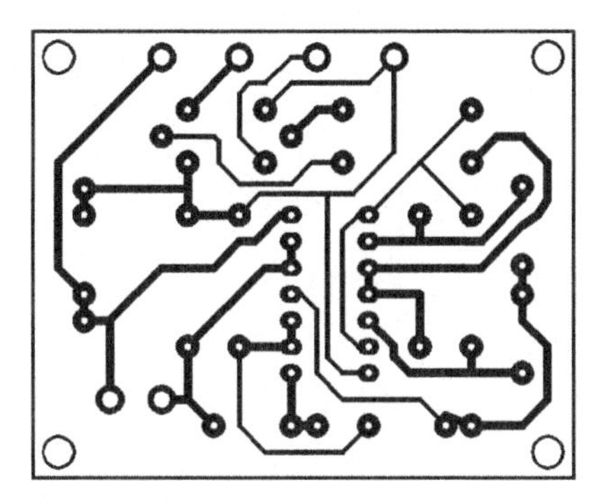

Figure 44.0 Printed Circuit Layout for Electronic Head or Tail

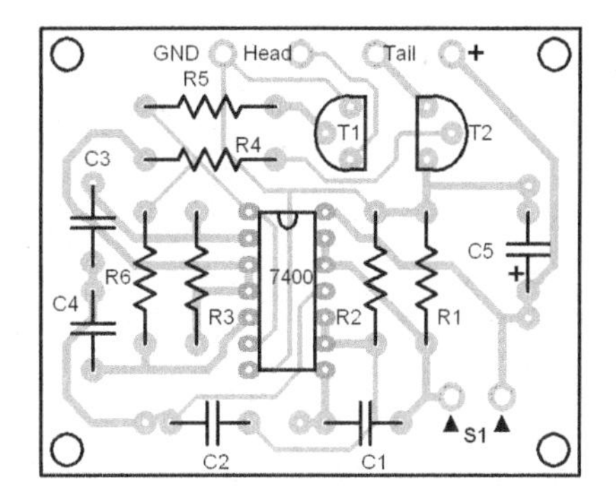

Figure 44.1 Parts Placement Layout for Electronic Head or Tail

45 POLARITY INDICATOR

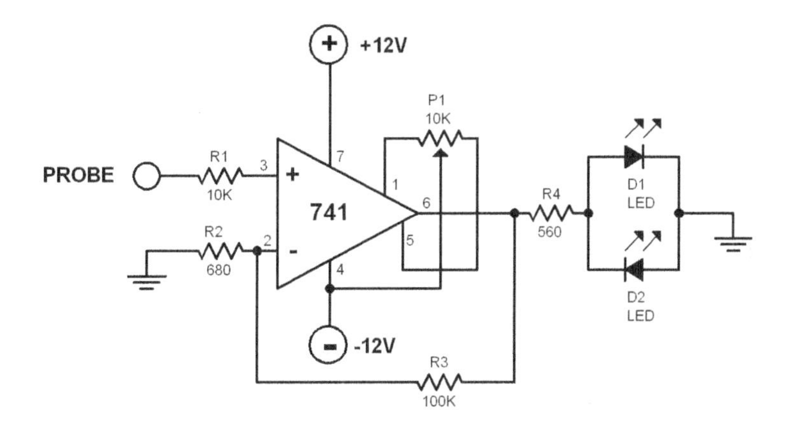

Diagram 45.0 Polarity Indicator

It is easy to test the polarity of a circuit's point whether it is positive or negative by using the tester circuit featured above. The tester circuit has a high input impedance (around 1 megaohms) to avoid loading the point being tested. In testing sensitive points however, (e.g. inputs of opamps), the input impedance must be taken into consideration.

The opamp 741 is the core of the tester circuit. Its non-inverting input is used to test the points for polarity. It has a gain of around 150 which enables it to test low voltage levels. The test result is displayed through the two LEDs D1 and D2. D1 lights up by positive polarity and D2 lights up by negative polarity. Take note that the pin profile on the diagram is based on the TO-5 package profile.

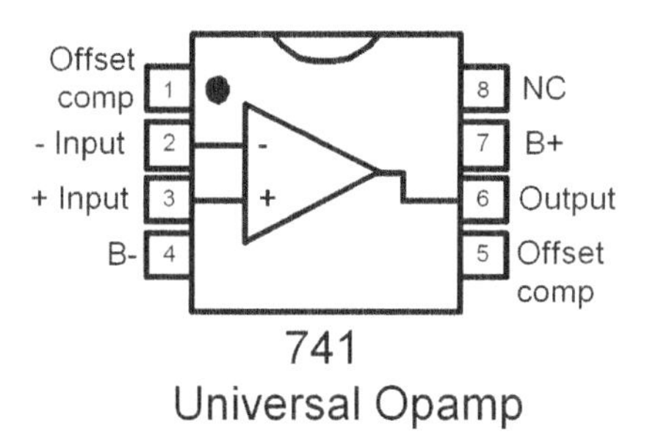

46 RIPPLE FILTER

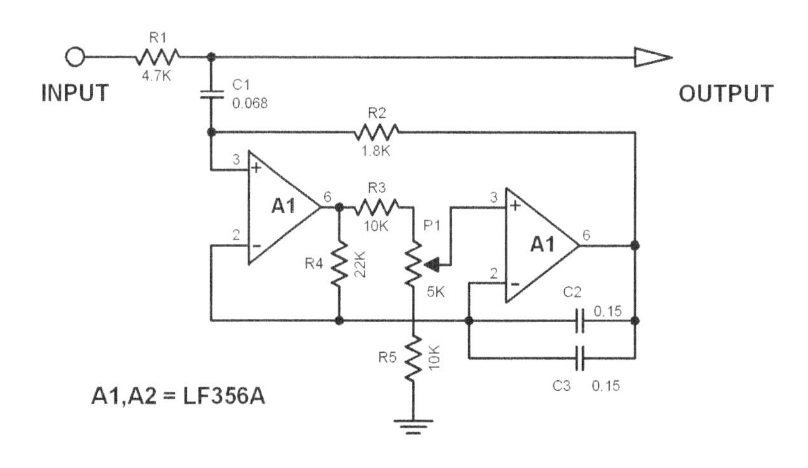

Diagram 46.0 Ripple Filter

This ripple filter is specially designed to filter out 50 Hz ripple signals from audio lines. A 50 Hz noise ripple is usually found in european countries. In many cases, it is not possible to remove the cause of the ripple noise. The featured active selective notch filter solves the problem in such cases. It allows the desirable signal to pass through with minimal attenuation. The Q factor of the filter is 10 at the inductivity value of 150 H.

These values can only be achieved practically by using an electronic „coil“ such as this circuit. The circuit works as an active RCL circuit. The two opamps A1 and A2 together with the resistors R2 to R5, C2 and P1, work as the electronic „coil“. The coils inductivity value is L = R2 x R3 x C1. Potentiometer P1 varies the inductivity value. The filter circuit attenuates the ripple noise by 45 to 50 dB, if properly adjusted.

47 SOUND EFFECTS CIRCUIT

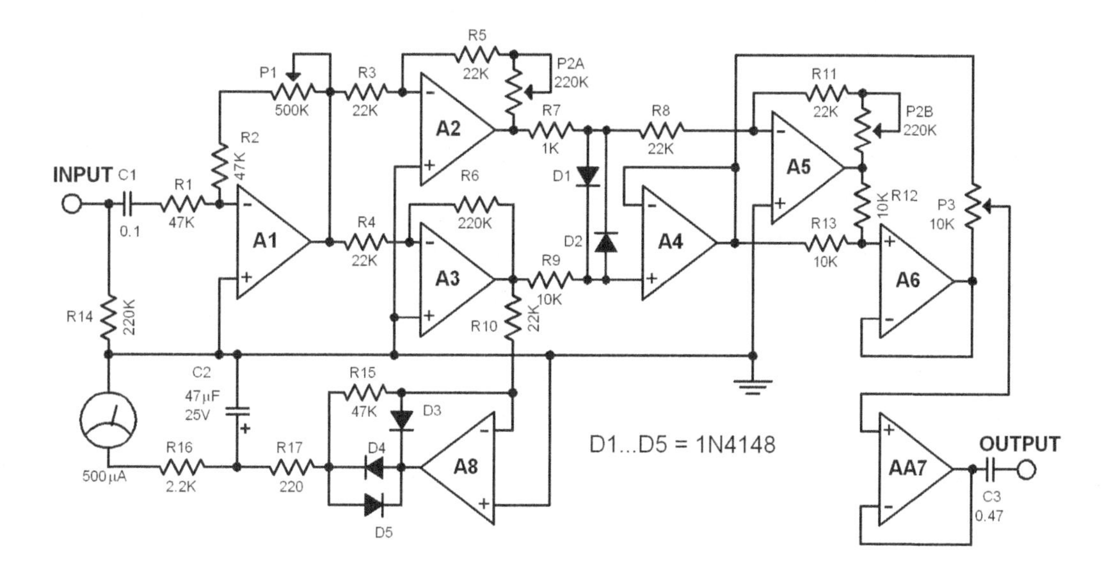

Diagram 47.0 Sound Effects Circuit

This circuit is designed to work as a signal „distorter“. If used with an electric guitar, it allows the production of special sound effects.The signal distortion is done by subtracting and mixing the normally clipped input signal with a processed „noise signals“. The outcomes are different waveforms. The variations can range from minimal to very strongly clipped signal up to a pulsed waveform which can have increasing dynamics. Potentiometer P3 adjusts the desired waveform while P2 adjusts the degree of distortion. Potentiometer P1 adjusts the input signal level which influences the output waveform.

The gain is dependent on the setting of P1 and the chosen effect degree. It is between 10 dB and 30 dB.

Opamp A1 is an impedance converter and signal buffer. Opamp A3 amplifies the signal constantly by a factor of 10 (20 dB). Diodes D1 and D2 clip the signal which is then fed to A4. The „noise signal“ comes from the junction of R9 and R10. Opamp A7 is another impedance converter. Opamp A8 is used to drive a VU meter.

How to use the sound effect generator: Set the input level with P1 until the VU meter points between 40 and 75% when a guitar string is struck. Select the distortion level with P2 from 0% to 100%. You can now select the waveform with P3.

48 Acupuncture Point Locator

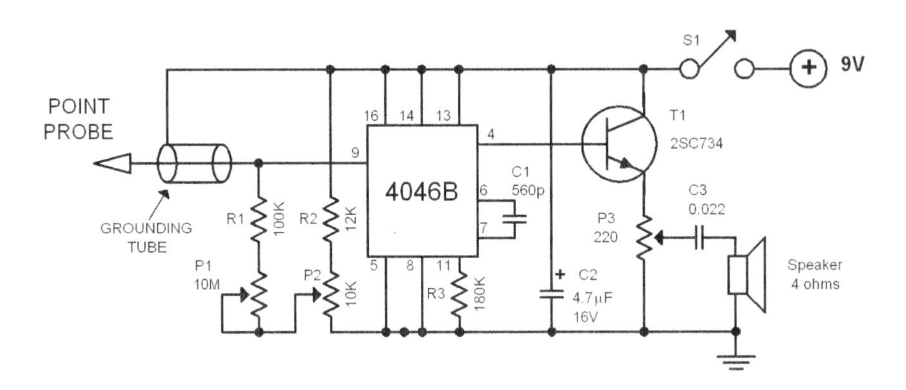

Diagram 48.0 Acupuncture Point Locator

This circuit is used to locate the sensitive acupuncture points on the human body. Acupuncture belongs to the traditional alternative medicine. This healing method requires the knowledge of the correct location of the important points on the body. Traditional acupuncture specialists rely on their expert knowledge to locate these points. Many of them also use electronic devices to help them locate these points. Most of these electronic devices measure the skin resistance based on the observed phenomenon that the skin resistance is lower on these acupuncture points.

The featured electronic circuit uses a digital PLL IC CD4046. This is a CMOS IC with an integrated VCO that is ideal for the application. The frequency range of the VCO is set by the RC circuit R3/C1. This is also dependent on a control voltage. This control voltage comes from a voltage divider circuit which is made of the skin resistance in conjunction with R1 and P1. With a decreasing skin resistance, the control voltage and the frequency increases. The signal can be heard through the speaker. Transistor T1 delivers enough power to drive the speaker.

Note:

Use only batteries to power the circuit. Never connect it to main power lines. The current consumption is around 45 mA at 9 volts.

49 ADJUSTABLE ZENER CIRCUIT

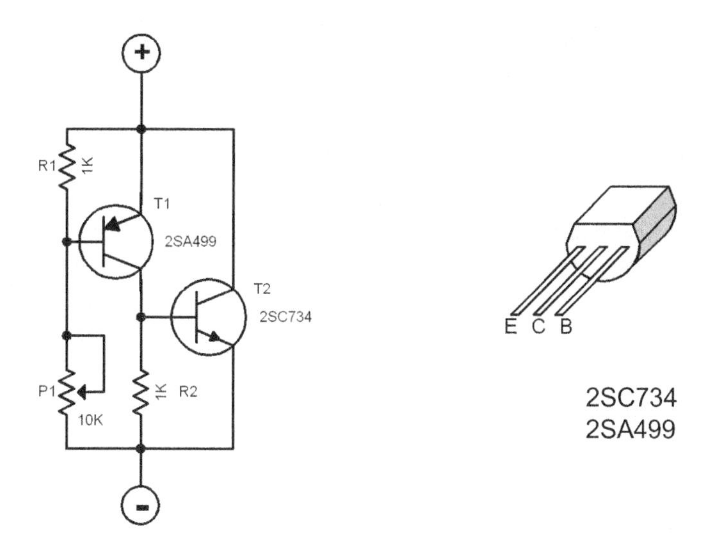

Diagram 49.0 Adjustable Zener Circuit

Zener diodes are widely used as voltage regulators. The voltage level that they regulate are usually given and fixed (called zener voltage). These values are based on the so-called E-12 series. However, if one needs a zener voltage that is not available on the E-12 series, this featured circuit comes very handy. This circuit can be considered a universal zener diode since its zener voltage can be varied.

The circuit uses two standard small transistors. It works starting at 0.8 volts and is ideal for zener voltage levels where there is no E-series equivalents. The given circuit will work up to a voltage of 6 volts. The voltage level can be set through P1 that is coupled with R1 as a voltage divider. When the voltage level is high enough for R1 to drop 0.6 volts, the transistor T1 starts to conduct. Its collector controls transistor T2. When the voltage at R1 and P1 increases, the transistor T2 conducts more thereby reversing the voltage increase. This is similar to a zener diode's characteristics. As an added bonus, the feature circuit is more stable than a fixed voltage zener diode. The circuit's internal resistance is around 3 ohms only.

The zener voltage of the circuit can be calculated in a simple way:

$$Uz = (I + P1/R1) \times 0.6V$$

For zener voltages higher than 6 volts, it is recommended to add one or two 1N4148 diodes in series with emitter of T1. The above formula must then be modified to:

$$Uz = (I + P1/R1) \times 1.2V \text{ (for 1 diode)}$$

or

$$Uz = (I + P1/R1) \times 1.8V \text{ (for 2 diodes).}$$

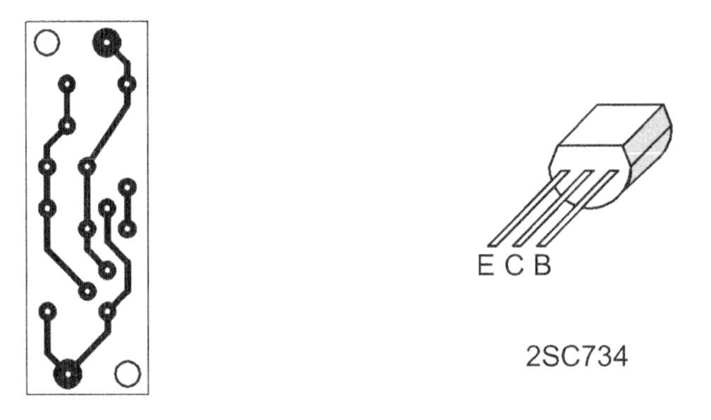

Figure 49.0 Printed Circuit Layout

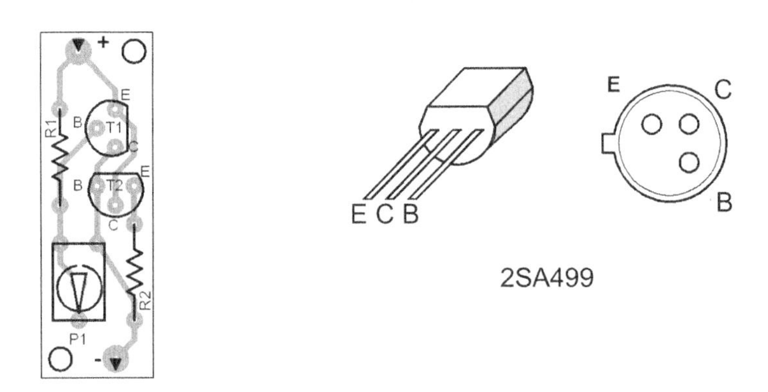

Figure 49.1 Parts Placement Layout

50 555 VOLTAGE DOUBLER

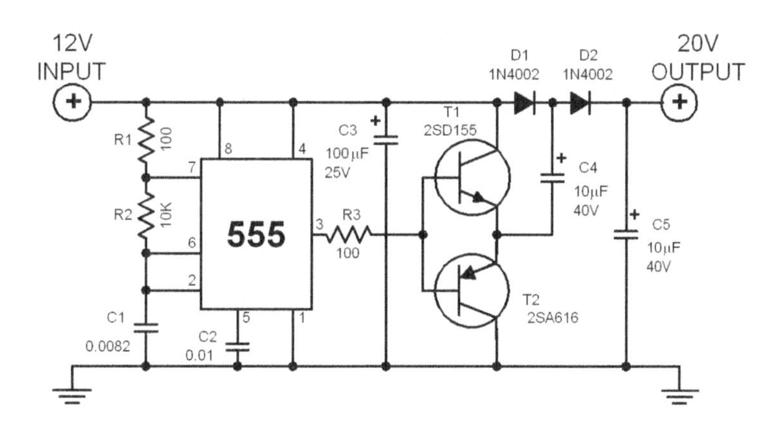

Diagram 50.0 555 Voltage Doubler

This circuit produces a voltage that is twice its voltage supply. This is useful when a higher voltage level is needed out of a single but lower voltage power supply. Since the current consumption levels are low in such cases, the circuit can be built with minimal resources.

The electronic circuit is basically a square wave generator using the common 555 timer IC. It is followed by a final stage made of transistors T1 and T2. The actual doubler circuit is made of D1,D2,C4, and C5 components.

The 555 timer IC works as an astable multivibrator and generates a frequency of about 8.5 kHz. The square wave output drives the final stage made of T1 and T2. This is how the doubler works: by a low amplitude of the signal, transistor T1 blocks while T2 conducts. The minus electrode of the capacitor C4 is grounded and charges through D1. By a high amplitude of the signal, transistor T1 conducts while T2 blocks. However, capacitor C4 cannot discharge because it is blocked by D1. The following capacitor C5 is therefore charged with a combined voltage from C4 and the power supply (12V input).

On standby, the circuit delivers around 20 volts. The maximum load must not exceed 70 mA. The actual output voltage is around 18 volts giving an efficiency rating of 32%. On lower current ratings, the voltage is higher. If a stable voltage level is desired, a 3 pin voltage regulator IC can be added at the output. The regulator IC's own current consumption must be added to the total current consumption which must not exceed 70 mA.

51 Trigger w/ Adjustable Threshold

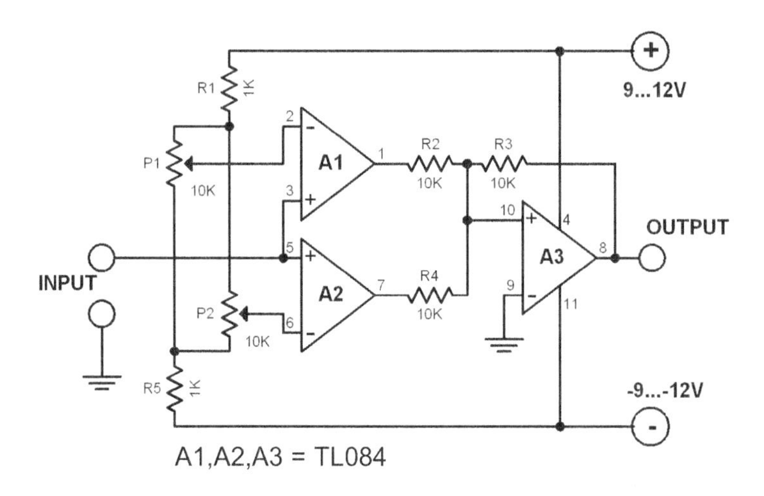

Diagram 51.0 Trigger w/ Adjustable Threshold

This trigger circuit offers an advantage of having an adjustable threshold. In most trigger circuits with switch hysteresis including the schmitt trigger, setting the threshold is very difficult since any change in the switching threshold affects the second threshold level. This problem can be avoided by using this featured circuit made of three opamps.

The two threshold voltages can be changed independently from each other through the potentiometers P1 and P2. Both positive and negative values within 0 up to 0.1% of the power supply.

When the input voltage is above the set upper threshold, the output of A1 and A2 will be „high" logic (positive voltage). The output of A3 is also positive. When the input voltage is below the set lower threshold, the outputs of A1 and A2 will be „low" logic (negative voltage). The output of A3 becomes negative.

This circuit can process both DC and AC voltages. Take note however that the maximum levels of the AC input signal must not exceed the power supply voltage.

52 RS Flip-flop w/ Inverters

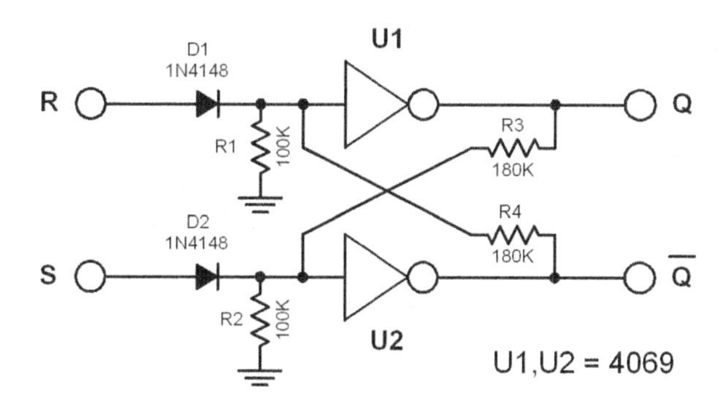

Diagram 52.0 RS Flip-flop w/ Inverters

RS flip-flops are usually built using NAND gates. Little known is the fact that they can be built also using inverter gates. The featured circuit shows an RS flip-flop built using CMOS inverter gates. The main differences to NAND gates are:

1. The inputs of inverters are at low („0") state at standby.
2. This requires the flip-flop to be set and reset with a positive pulse („1").
3. Both outputs Q and (-)Q are inverted as opposed to a NAND RS flip-flop.

A positive pulse („1") at S-input sets the lower inverter output to „0"; this „0" goes to the input of the upper inverter via R4. As a result, the Q output becomes „1" (high). This „1" goes to the input of the lower inverter via R3 thereby sustaining the state of the flip-flop even after the pulse is removed from the S-input.

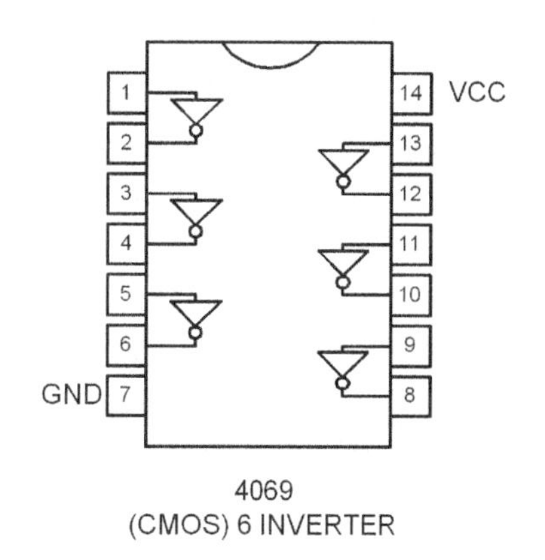

4069
(CMOS) 6 INVERTER

53 DC REGULATOR

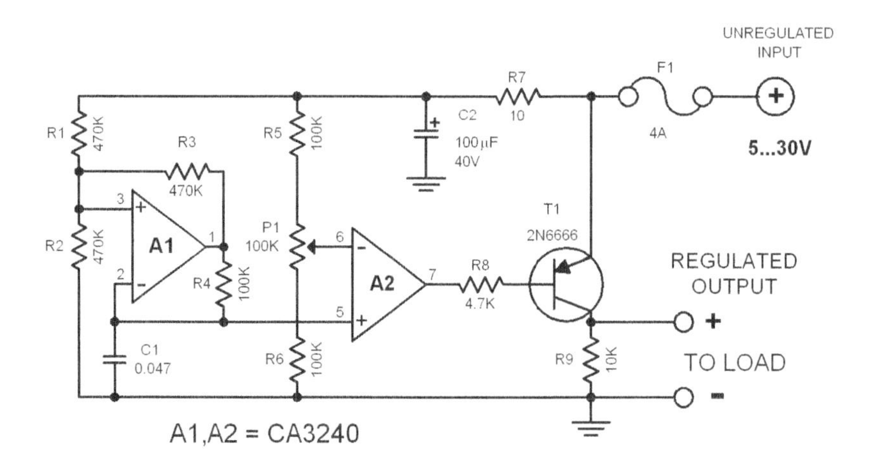

Diagram 53.0 DC Regulator

This circuit regulates a DC power output. It has a very wide application range. It can be used to control the speed of a motor, a pump, a toy train, the brightness of a LED or lamp, etc. Practically, it can be used in any application that uses a regulated DC power with pulse width modulation (PWM). It was originally used as a jumbo LED dimmer.

The circuit works this way: the A1 opamp functions as a square wave generator. At its non-inverting input is a by product triangle wave signal (almost triangle anyway). The IC A2 following it functions as a simple comparator. A reference voltage is fed to the inverting input of the A2 IC via the potentiometer P1. The output of A2 is a square wave signal with a constant frequency of around 200 Hertz. This signal has a variable pulse width from 0 to 100%. The P1 sets the trigger point of the pulse. The transistor T1 works as the actual regulator by switching a relatively high current with a maximum of 5 amperes. The power supply voltage must be between 5 volts (minimum) and 30 volts (maximum). Take note that the lower the supply voltage is, the lower the efficiency of the circuit becomes.

54 3-AMPERE POWER SUPPLY

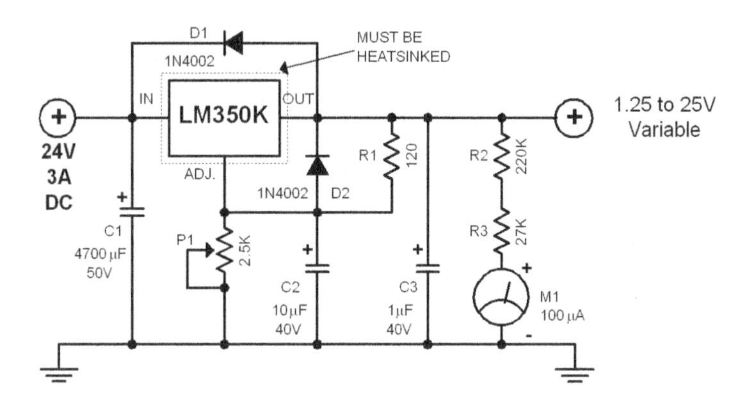

Diagram 54.0 3-Ampere Power Supply

This is a compact power supply that delivers a stable voltage and currents up to 3 amperes. The circuit is very conventional and its voltage output can be varied from 1.25 volts to 25 volts.

The main module is the LM350 IC which integrates a voltage regulator and a power stage. It also has a built-in overload protection which activates at 30 watts of power dissipation. The voltage output is set by connecting the „adj" pin of the IC to the voltage divider made of R1 and P1. The output voltage can be calculated using the following formula:

$$1.25\ V \times (1 + P1/R1)$$

where the P1 value is between 0 and 2.5 kiloohm. Capacitor C1 is a common ripple filter while capacitors C2 and C3 improves the regulation. The diodes D1 and D2 serves as protection for the regulator IC when the IC output is turned off. Resistor R1 is 120 ohm. This ensures that the minimal load current for the IC (around 3.4 mA) is high enough to maintain good performance.

One thing that is most important in building the electronic circuit: provide adequate heatsink for the LM350 IC. The power dissipation at the IC is very high, around 85 watts.

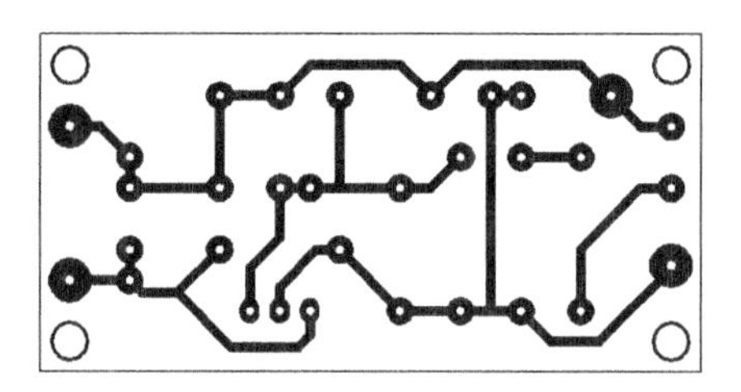

Figure 54.0 Printed Circuit Layout

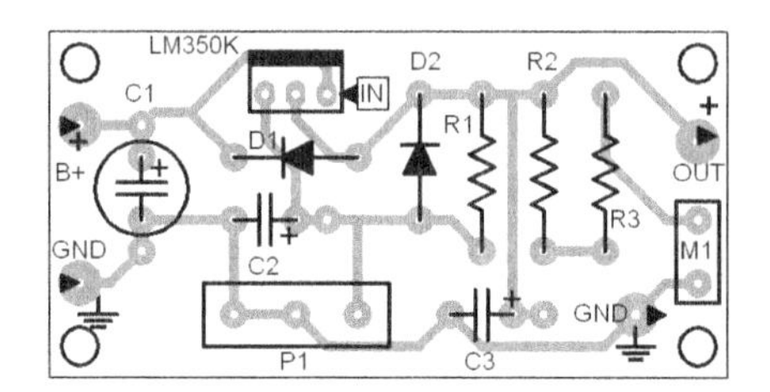

Figure 54.1 Parts Placement Layout

Consider that the heat resistance of a TO-3 package is 1.5°C/W and the maximum allowable temperature is 150°C. If a heatsink is used with a heat resistance of 1.5°C/W, then the total heat resistance is 4°C/W. At 30 watts dissipation and 25°C outside temperature, the resulting internal IC temperature is 145°C. Once this dissipation level is reached, the internal protection activates shutting down the IC.

One way to avoid high dissipation levels is to use a lower voltage transformers when the needed output voltages are low. To put it simply: if you are using the circuit to supply voltages around 9 volts, do not use a 25 volts transformer but use a lower voltage one instead (e.g. 12 or 15 volts).

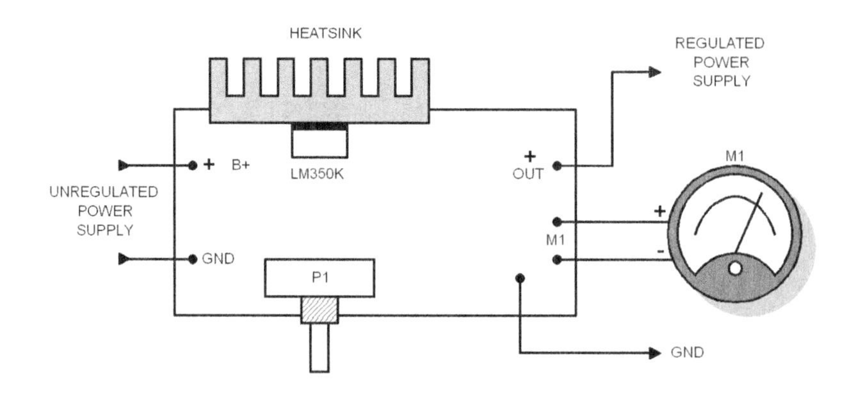

Figure 54.2 External Wiring and Heatsink Layout

55 ADJUSTABLE CAPACITANCE

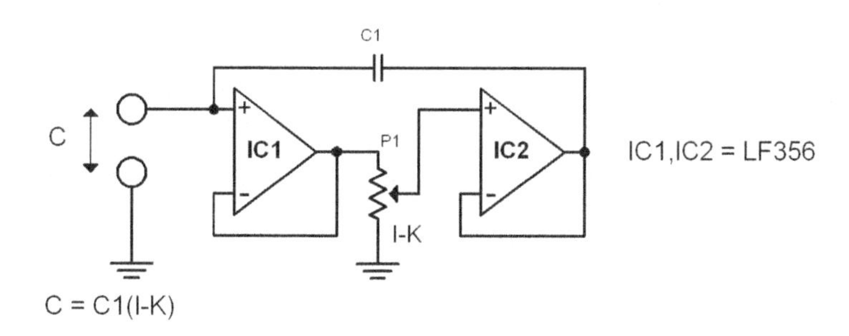

Diagram 55.0 Adjustable Capacitance

The featured adjustable capacitance circuit is made up of two opamps configured as voltage follower. The two opamps build up the capacitance with a capacitor C1 and made adjustable with a potentiometer P1. P1 works with C1 and some capacitance called parasitic capacitance. The electrodes of this capacitance is shown in the diagram as C (plus input of IC1 and the ground). Since the IC1 is a voltage follower, the input potential also appears at P1. Part of this potential labeled kUi is fed to IC2. A factor k represents the partial resistance in relation with total resistance P1. A voltage (1-k)Ui is present at C1. The capacitive charge is therefore equivalent to : $Q = CU$, or $Q=Ci(1-k)Ui$.

The effective capacitance is :

$$C=Q/Ui = C1(1-k)Ui / Ui$$

or simply

$$C = C1(1-k).$$

Pin	Function	Pin	Function
1	Null Adjust	8	NC
2	(-) Input	7	B+
3	(+) Input	6	Output
4	B-	5	Null Adjust

LF356

56 CAPACITIVE TOUCH SWITCH

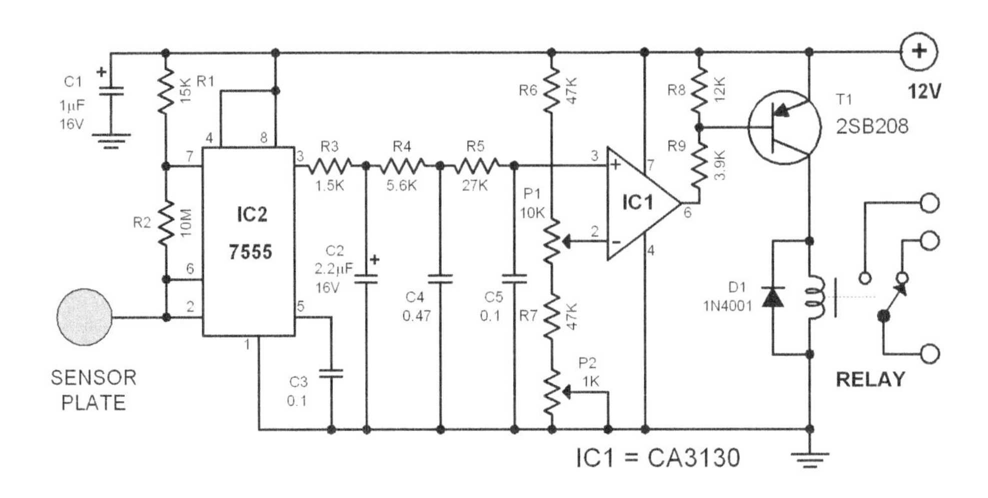

Diagram 56.0 Capacitive Touch Switch

This circuit uses the following trick: if a square wave signal with a 50% duty cycle is integrated, it results to an stable averaged voltage. If the signal frequency is changed the integrated voltage will remain constant. However, positive negative voltages spikes will appear due to quick changes in the duty cycle. These „spikes“ are being put to good use in this circuit.

In the circuit, the internal capacitance of the 555 timer enables the square wave oscillator configuration to oscillate. If a capacitive sensor is added to this oscillator, it is possible to change the oscillation frequency simply by getting closer or farther to the sensor. The resulting square wave signal is integrated by an RC network (R1,R2, sensor plate). The voltage spikes are fed to the IC1. IC1 is a comparator with variable reference. If one gets close to the sensor plate, the relay activates. When one goes away from the sensor plate, the relay releases. When the sensor is touched, the capacitance value of the plate increases and the oscillator frequency decreases. This produces negative going voltage spikes.

In constructing this electronic circuit, make sure that the sensor plate is dimensioned so that the square wave generator oscillates at around few kilohertz. Lower frequencies would make the circuit prone to false alarms. Potentiometer P1 is for setting the comparator threshold while P2 is for fine adjustment.

57 ANALOG-DIGITAL CONVERTER

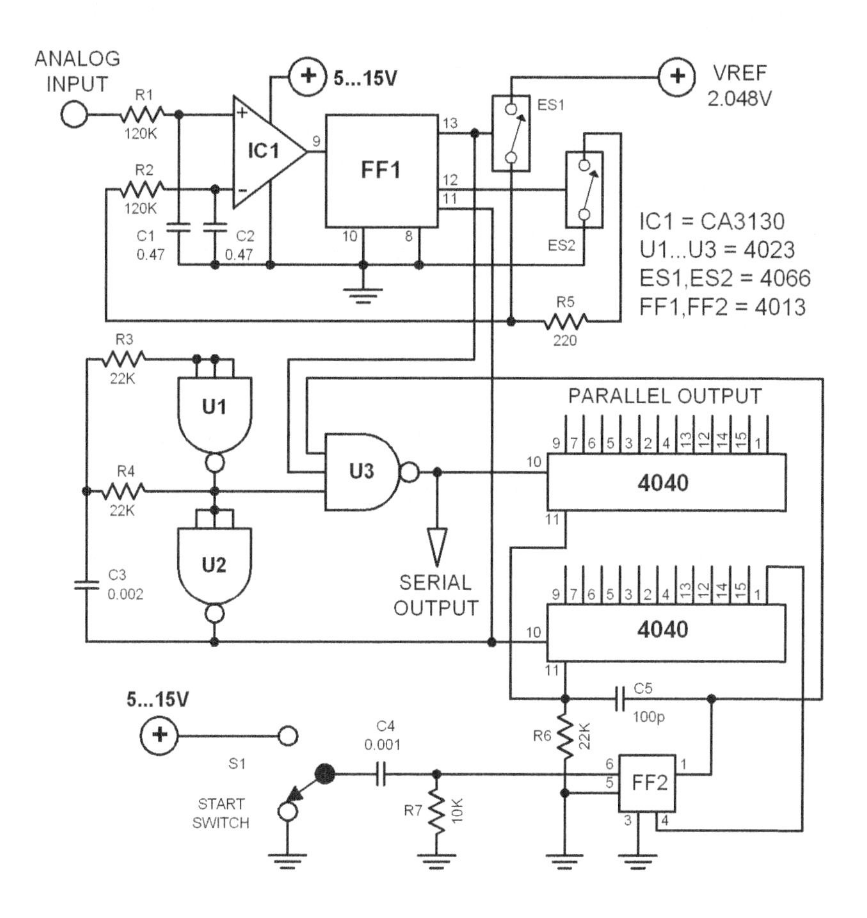

Diagram 57.0 Analog - Digital Converter

Analog to digital converters belong to special circuits since they usually require special electronic components and careful calibration. The featured circuit however works under the delta-sigma principle hence it does not need special components. Its accuracy depends entirely on the stability of the external reference voltage. The two integrated circuits IC1 and IC2 work as a delta-sigma (or delta) modulator in conjunction with the two CMOS switches S1 and S2 .

The inverting input of opamp IC1 gets switched alternately to the ground or to the voltage reference Uref via R2 and one of the two CMOS switches. This switching action is dependent on the voltage at the D-input of the flipflop FF1. The IC1 works as a comparator opamp. The feedback to the non-inverting input switches IC1 back and forth causing the average values of both opamp input voltages to balance out. As a result of this, a squarewave signal comes out of the ICD output. Its duty cycle is proportional to the level of the input voltage. The switch mode of the comparator is controlled by the clock impulses from the flipflop FF1.

When FF1's Q output is „1", clock pulses get through to the input of IC5 (a 12 bit counter) via gate U3. The period length of the counter inside IC5 is controlled by a second binary counter 4040. This counter counts down to 218 then delivers a reset pulse to flipflop FF2. This results to a „0" at its Q output, U3 blocks and IC5 does not receive any control pulses. The ending count at the second counter 4040 therefore represents the level of the analog input voltage. The next conversion period starts when the „start" button at FF2 is pressed. This resets the IC5 and IC6 at zero position.

If the reference voltage is exactly 2.048 volts, the analog voltage input of 1 volt will result to 1000 counter pulses. The components used in the circuit must have tolerances of less than 1%. If you can use a symmetrical power supply, you can achieve much better stability by using LF357 as the comparator IC. The clock frequency of the oscillator (U1, U2) is 8 kHz. It is dimensioned through R3,R4 and C3. Decreasing the value of C3 will increase the clock frequency. However, C3 cannot be lower than 39 pF. A 39 pF for C3 will generate a clock frequency of about 50 kHz. Other options are: using a different output pin of IC6 (4040) or using a different counter type for IC5 and IC6.

The calibration of the circuit is limited to compensating the offset bias of IC1. Short the inputs and adjust P1 until the counter IC5 stands at zero. The current consumption of the circuit is just several milliamperes. This A/D circuit is most effectively used in conjunction to a microcomputer that processes the final data.

The capacitor C6 is not shown in the schematic diagram but shown in the parts placement layout. It is a tantalum type and rated 1µF/25V.

The potentiometer P1 works as the offset trimmer, rated 100K and connected to the IC1 as shown in diagram 57.1.

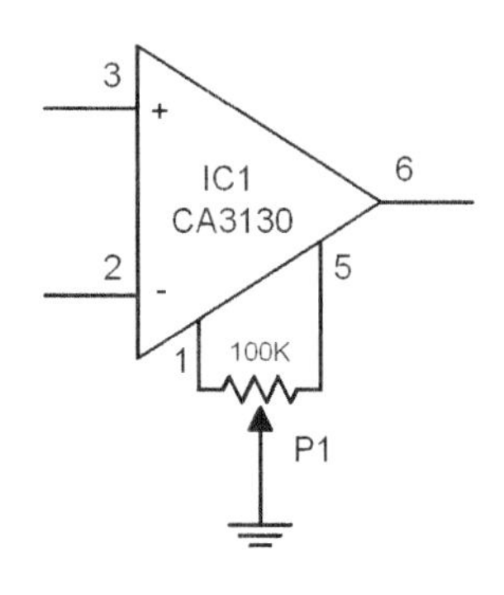

Diagram 57.1 Offset trimmer P1

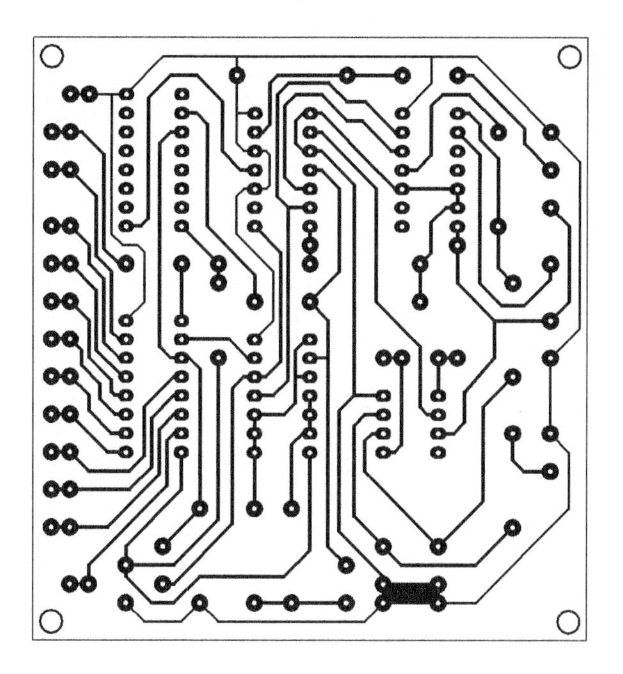

Figure 57.0 Printed Circuit Layout

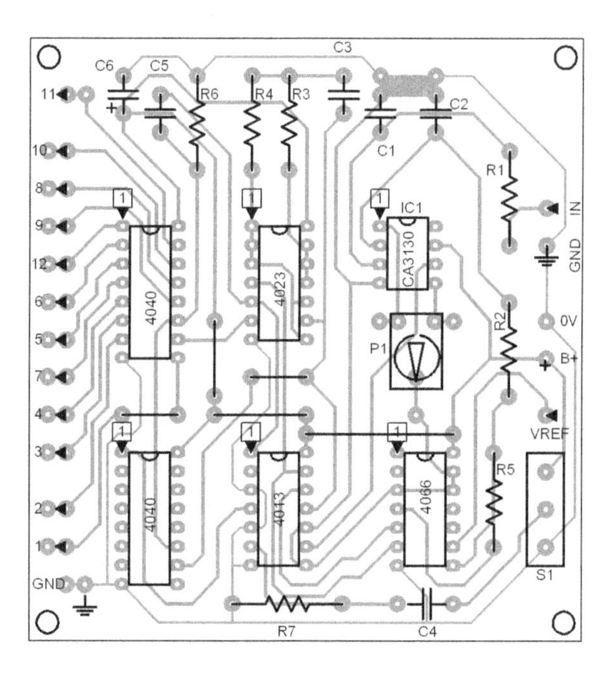

Figure 57.1 Parts Placement Layout

58 Moisture Controlled Switch

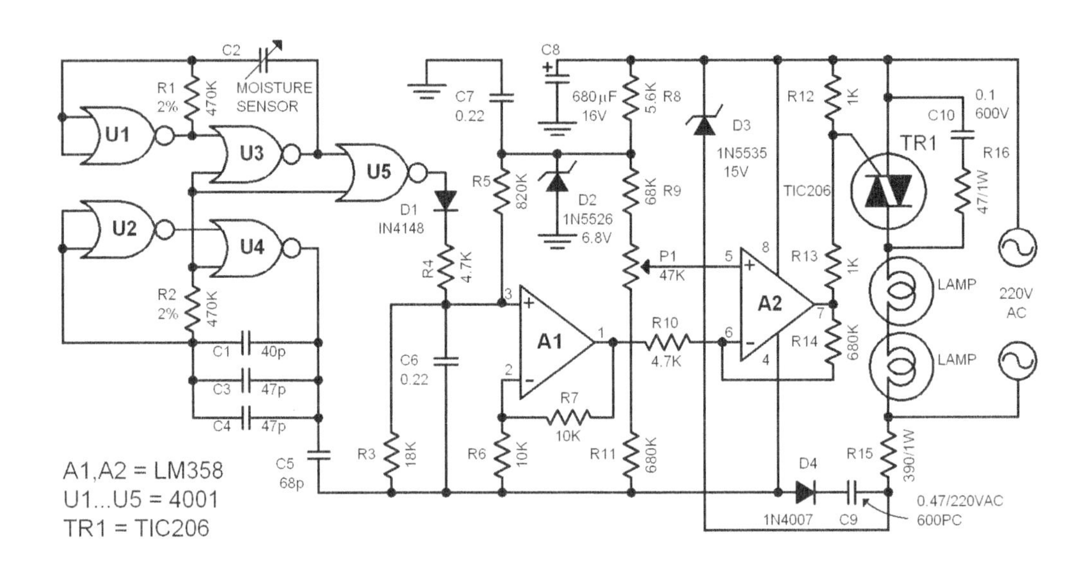

Diagram 58.0 Moisture Controlled Switch

This circuit turns on and off an electrical load (originally intended to switch a heater) depending on the moisture content of the surrounding air. The moisture is „sensed" with the help of a plate capacitor C2. This capacitor is similar to the old air dielectric plate capacitors in your vintage AM radio used to select the radio stations. At a certain moisture level, the circuit switches on the load (a heater). The moisture level is converted to a voltage through R3 in the first part of the circuit. The first opamp A1 in the circuit functions as a high impedance. The second opamp A2 functions as a comparator with a hysteresis of about 15%. The variable scale of the potentiometer P1 produces voltages from 0.6V to 3 volts and proportional to moisture levels from 20 to 100%. The P1 therefore sets the moisture level where the circuit activates. Once the moisture level reaches above the set level (via P1), the triac TR1 triggers and the attached load (heater) turns on.

The current consumption of the circuit is around 13 mA while activated. If you use it to control a heater, you can plug in two 100 watts bulbs. If you use light bulbs as heating element, enclose them in a metal container.

Calibration: Dissolve some salt in a glass of water and put it inside a closed room. Soon, the moisture level will climb to about 75%. Then set C2 to get a voltage of 2.25 volts at R3. After that, set P1 to a level that triggers the triac. This concludes the calibration. Later, when you use the circuit, it will trigger at around 80% air moisture.

59 FUSE & LAMP TESTER

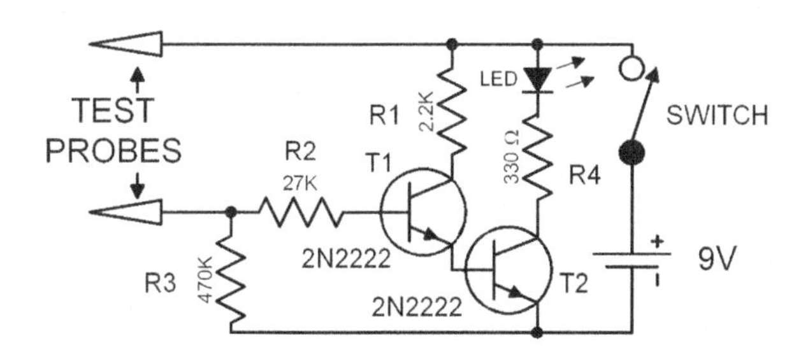

Diagram 59.0 Fuse and Lamp Tester

Testing cables, wires, fuses, lamps, etc. belongs to a repair job like butter to a bread. Sometimes, testing parts becomes too cumbersome since a person has only two hands and too often, one has to hold the part being tested and the two probes of an ordinary continuity tester simultaneously. It wouldn't be much of a problem if only one had an assistant. A third hand? Well, you never have seen a mutant repairman yet, have you?

This electronic circuit enables easy testing of lamps and fuses by using the conductivity of the human body. One of the test probes is connected to the part under test while the other probe is held by the normal hand (bare, of course). When the lamp or fuse is working properly, a small amount of curent flows through the hand which is enough to switch on the transistors and light the LED.

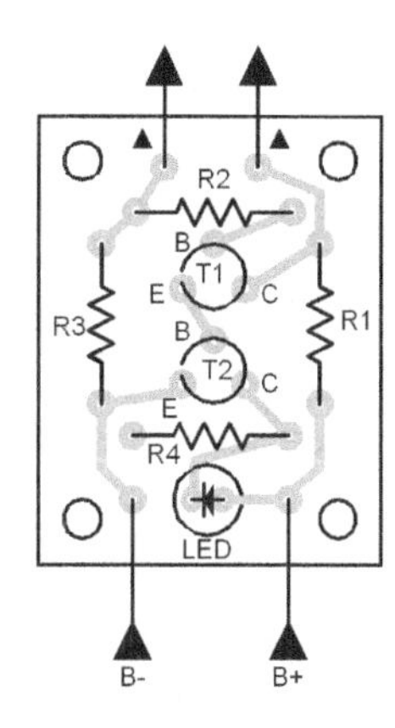

Figure 59.0
Parts Placement Layout

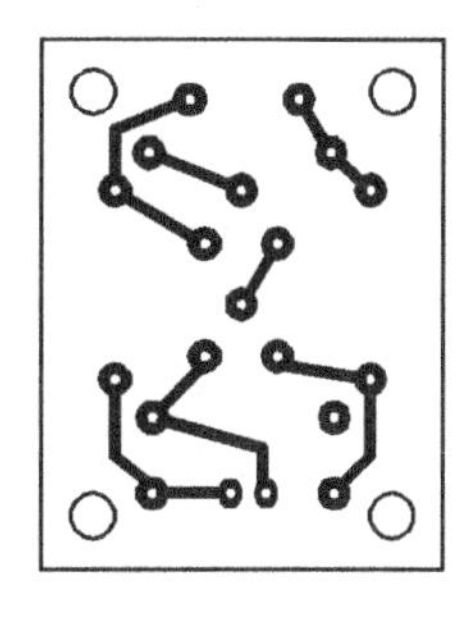

Figure 59.1
Printed Circuit Layout

60 LOW CURRENT RELAY

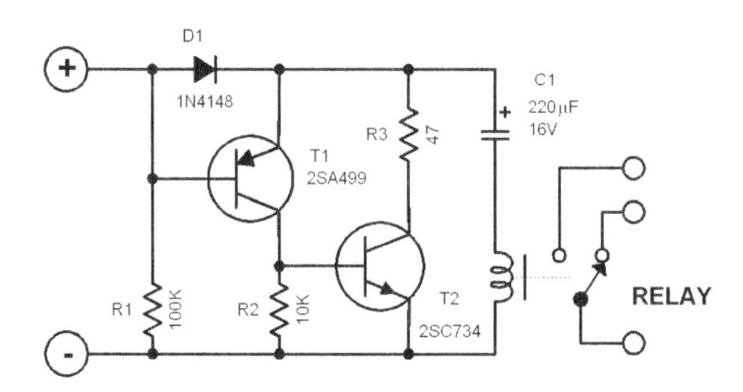

Diagram 60.0 Low Current Relay

This circuit is designed to be used in battery operated electronic devices. Its special design enables it to consume neglegible operating currents. Its operating current is in micro amperes! The trick is done by using a bistable relay and adding some non-critical components to force the relay to behave like a monostable relay. A bistable relay stays at its last state when the power is turned off. However, most bistable relays consume at least 50 mA trigger current. A monostable relay however switches back to its original state when the power is turned off.

The electronic circuit works this way: when the power is turned on, the capacitor C1 charges via D1 and the relay coil. This charging current activates the relay. Diode D1 ensures that the base of T1 is always more positive than its emitter. T1 and T2 are in this case always blocked. Once the power is turned off, the emitter of T1 is coupled to the charge voltage at the positive pole of C1. Its base and the relay coil on the other hand are coupled to the negative pole of C1. Transistor T1 and T2 conduct. Now, the capacitor C1 can discharge through T2 and relay. The current flows to the relay coil but in reverse direction. The relay is activated to its other state. The bistable relay with the added components behave like a monostable relay. It has the advantage of consuming very little current. The current is around 150 microAmperes! For a reliable operation, select the relay's operating voltage as 2/3 to 3/4 of the main power supply. For example: for a 12 volt power suppy, select a 9 volt relay.

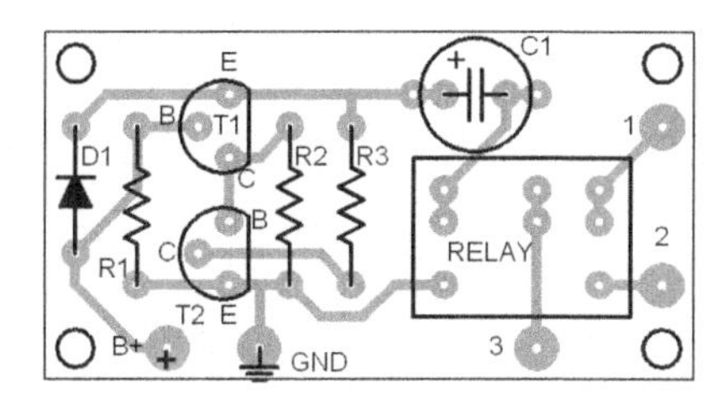

Figure 60.0 Parts Placement Layout for Low Current Relay

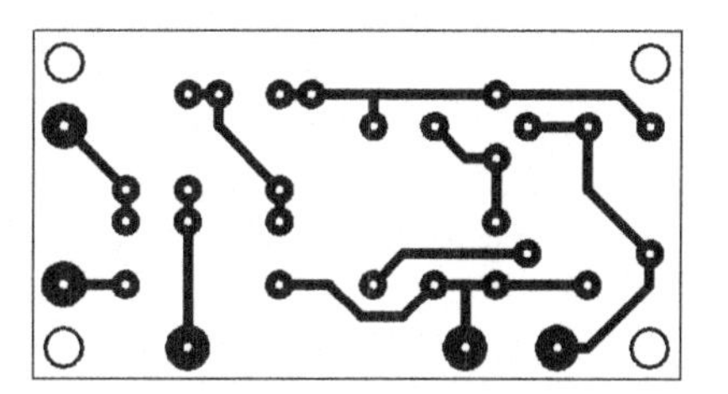

Figure 60.1 Printed Circuit Layout for Low Current Relay

61 Automatic Park Light Switch

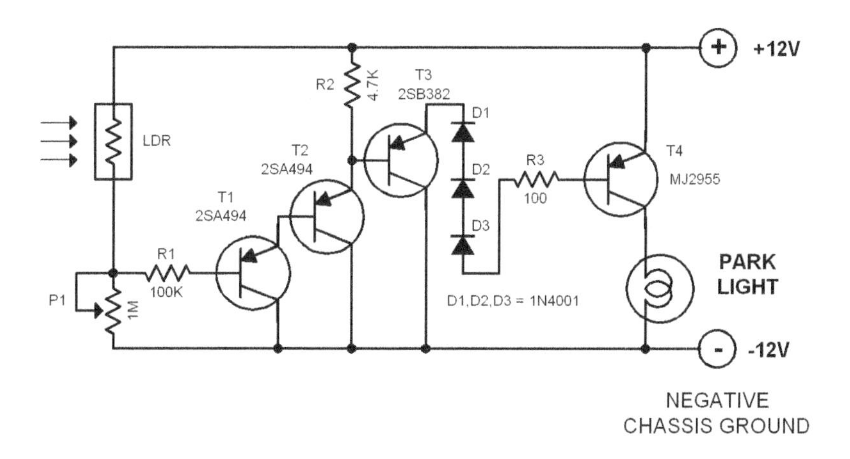

Diagram 61.0 Automatic Park Light Switch (negative chassis)

This circuit enables the car's park light to automatically turn on when the surrounding light dims to a preset level. The circuit shown in diagram 61.0 is an NPN design intended for negative grounded chassis. The diagram 61.1 is the PNP circuit intended for a positive grounded chassis. The dim level at which the circuit activates is set through the potentiometer P1.

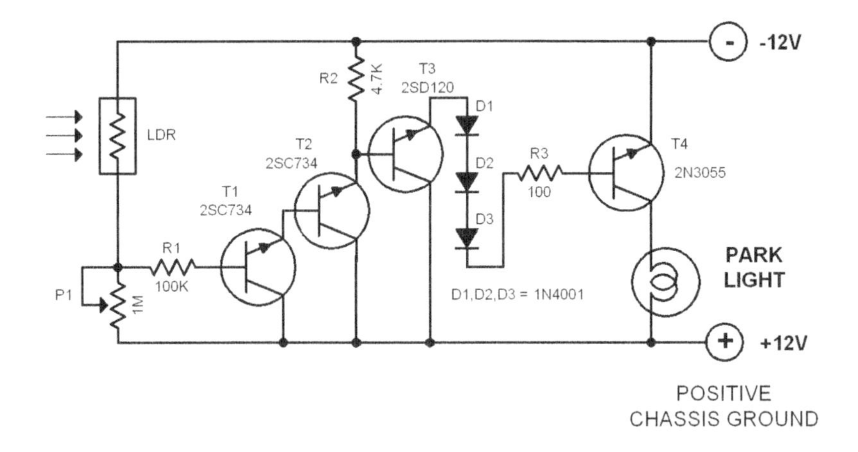

Diagram 61.1 Automatic Park Light Switch (positive chassis)

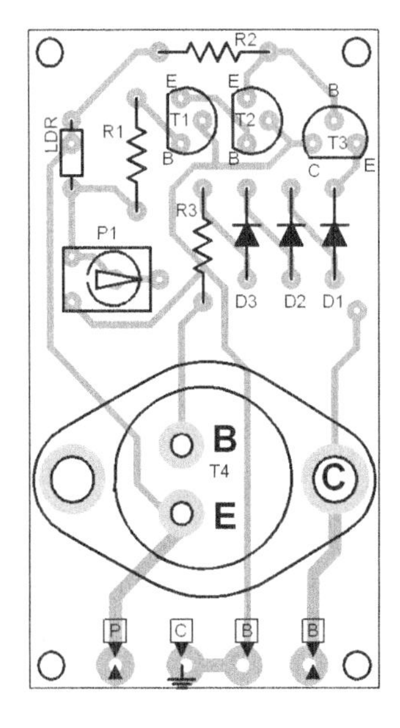

Figure 61.0 Parts Placement

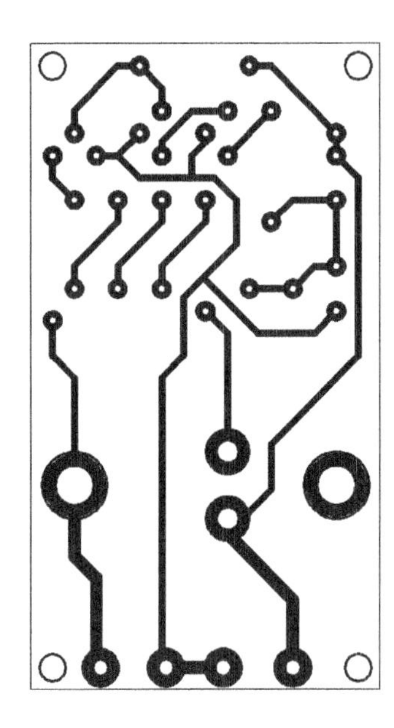

Figure 61.1 Printed Circuit

The printed circuit board layout shown in Figure 61.0 can be used for both the negative and positive polarity chassis. Take note however that the transistors are different for each chassis type. The connectors labeled as shown in Figure 61.2 have the following connections:

P = Power line of the car
C = Car chassis
B = Park light bulb

62 WHISTLE PROCESSOR

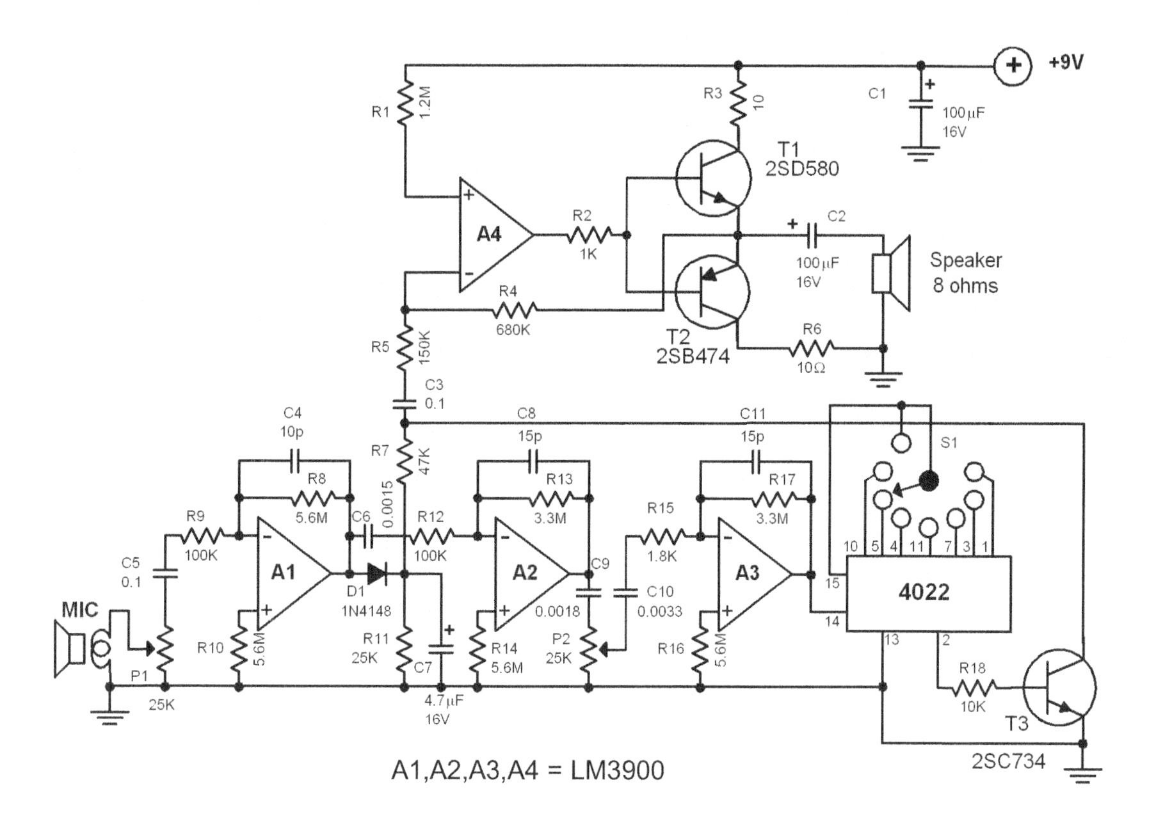

Diagram 62.0 Whistle Processor

This circuit processes the whistle into a tone with lower frequency but with almost the same amplitude as the original tone. The resulting tone is lower but with the same dynamics. The microphone being used is of the crystal type. The switch S1 selects the point where the counter will be reset. This way, the divider factor from 1 up to 8 can be selected.

The circuit can be used as an octave divider with an electric guitar.

63 POLARITY PROTECTION

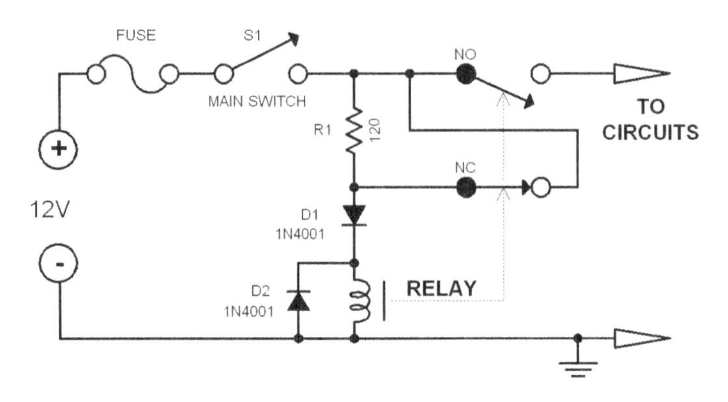

Diagram 63.0 Polarity Protection

Most electronic devices that are powered externally from a power supply are at risk of being powered with reverse polarity voltage. This could damage the device. The most simple protection technique is to connect a series diode to the power line input. The diode conducts only when the power supply polarity is correct. By reversed polarity, the diode blocks the power line. This however works well only at low current consumption levels. At higher current levels, the voltage drop and power loss at the diode affects the power level adversely.

The featured electronic circuit avoids the voltage and power loss problem. The circuit shown in diagram 63.0 is dimensioned for 12 volts power supplies. It is built into the electronic device to be protected. By correct polarity, the current flowing to the diode D1 and the relay coil causes the relay contacts to activate. The NO contact closes allowing the power to go to the electronic device. The NC contact opens and the current supplying the relay coil is reduced to a low level just enough to maintain relay activation. R1 reduces the current to the relay to keep a minimum power loss.

By reversed polarity, the diode blocks the current and the relay cannot activate. The NO contact remains open and the electronic device is protected from the reversed polarity voltage.

64 Autocharging Current Monitor

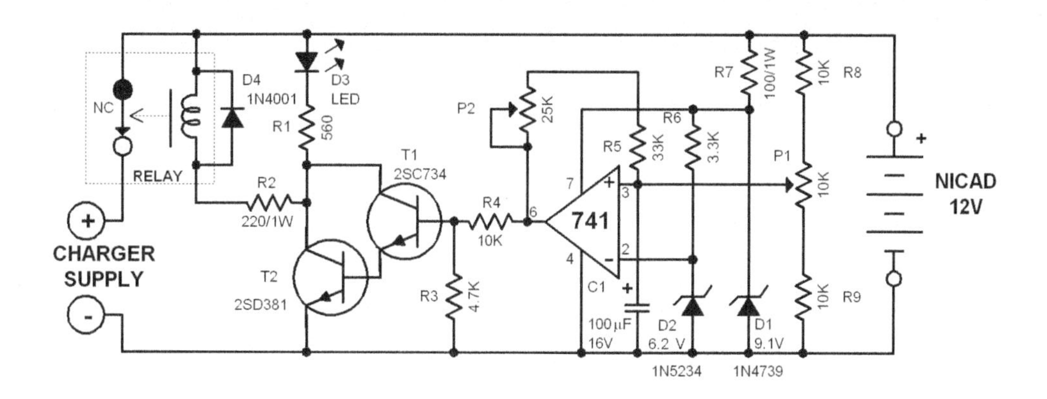

Diagram 64.0 Autocharging Current Monitor

This circuit is presented here as an alternative to expensive automatic battery chargers. The trick to lower the cost is to buy a cheap battery charger and expand it with an automatic charge breaker circuit such as the one featured above. The heart of this electronic circuit is the comparator. It compares the NiCad battery's voltage with a reference voltage. When the NiCad battery's voltage exceeds a certain presettable maximum voltage level, the circuit breaks the charging via the relay. On the other hand, when the NiCad battery's voltage sinks below the preset mimimum level, the circuit closes the relay and the charging resumes.

The 741 IC is being used as the comparator in this circuit. The power supply to the IC is being stabilized by R7 and D1 to make it independent from the battery voltage level. From this stabilized voltage, the reference voltage is taken via the R6 and D2 and fed to the inverting input of the comparator IC. The battery voltage is reduced by the R8, P1 and R9 divider network and compared with the reference voltage. When the battery voltage level rises, the voltage at the non-inverting input will also rise and by a certain level (set with P1) the non-inverting input has a higher level than the inverting input. This triggers the comparator to increase its output voltage thus driving the T1/T2 transistors to opening the relay (the relay is normally closed). This breaks the charging current. At the same time, an LED (D3) lights up signalling that charging is done.

The comparator is designed to have a hysteresis that prevents the circuit from „oscillating" (constantly switching on or off the relay) that could be caused by slight changes to the battery voltage. This hysteresis is set with the potentiometer P2. The setting of the hysteresis with P2 also sets the minimum battery voltage level at which the charging resumes. Calibrating the circuit: The best way to calibrate the circuit is to use a variable voltage, regulated power supply. This is connected to the circuit in the place of the NiCad battery. Set the regulated power supply to 14.5 volts and adjust potentiometer P1 to a point where the relay snaps open. Next, set the regulated power supply to 12.4 volts and adjust P2 to a point where the relay closes back. You might need to repeat this procedure severals time since P1 and P2 have an effect on each other.

65 6V to 12V CONVERTER

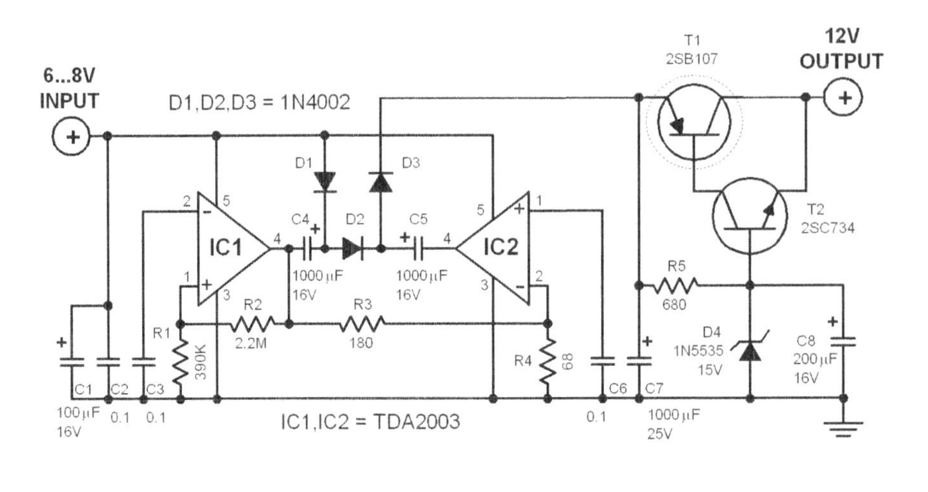

Diagram 65.0 6V to 12V Converter

In some situations, one needs a 12 volts power supply but has only a 6 volts battery available. In such a case, a voltage converter such as this featured circuit comes very handy. This circuit is a simple converter made with an IC from SGS with several additional external components. The IC is a TDA2003 but it can be replaced with a TDA2002. The cost of building the circuit should be low enough to justify constructing it instead of modifying the entire equipment setup to work directly with a 6 volts power supply. The two principles of simplicity and low cost are applied in this electronic circuit. It only uses two low cost AF amplifiers and functions properly without the need for a transformer.

The IC1 opamp functions as a stable power multivibrator. Its oscillation frequency is determined by the capacitor C3. Its oscillates at around 4 kHz at standby and increases in a loaded condition up to around 7 kHz. The output of the IC2 opamp is identical to the IC1 oscillator signal but in the opposite phase.

When the output of IC1 is at zero (saturation point of the output transistor to the ground), the capacitor C4 charges via the diode D1 up to the power supply level minus the voltage drop at D1. When the IC1 swings to the opposite direction, its output becomes positive. The output voltage from IC1 adds up to the voltage stored at the capacitor C4 forcing the diode D1 to stop conducting. Capacitor C5 then charges via the diode D2 to a voltage that is double than the power supply level. Due to the opposite phase of IC2, the negative pole of C5 is connected to the ground through the IC2 at this moment. By the next swing of IC1, its output is again at zero potential. At this moment the capacitor C4 is recharged while at the same time the voltage at C5 is „lifted“ up to the voltage level present at IC2.

Capacitor C5 delivers its voltage to the output capacitor C7 via the diode D3. Theoritically, the output voltage has been multiplied by three. In practical application, however, the output voltage available from the capacitor C7 is lower than the theoritical value. This voltage level is also dependent on the load. Data gathered from testing the circuit showed that a 6 volts supply (from a battery) produces a voltage output of around 18 volts with no load to the circuit. This level sinks to about 12 volts when a load is applied that consumes a current of around 750 milliamperes. When the load current is about 400 mA, the voltage at the capacitor C7 is about 14 volts. These values are enough to power a standard car radio.

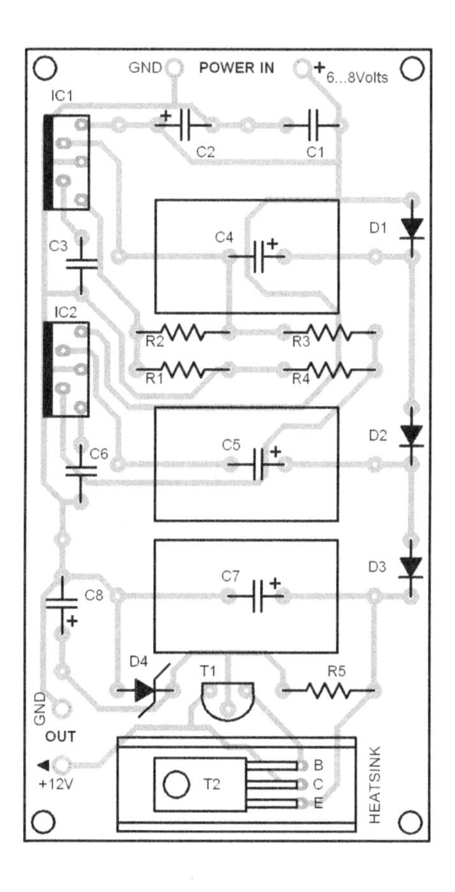

Figure 65.0 Parts Placement Layout

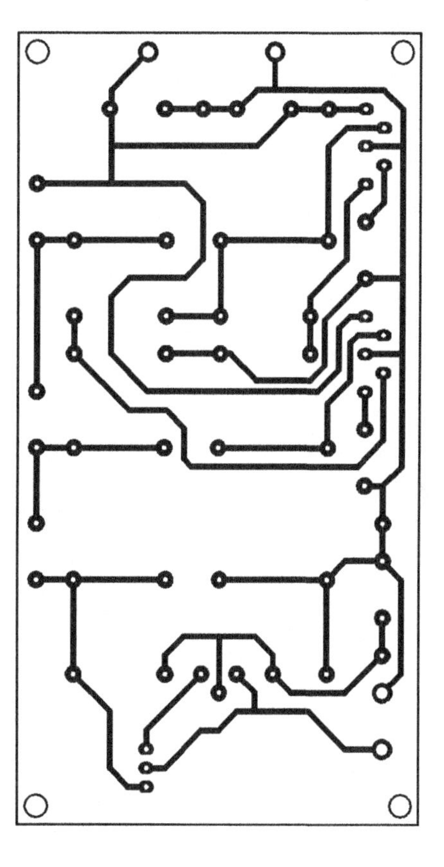

Figure 65.1 Printed Circuit Layout

As mentioned above, the theoritical output could reach the triple of the supply voltage. To guard against unnecessary voltage increases at low current consumptions, a limiter stage was added to the circuit composed of a 15 volt zener diode and a darlington transistor T1/T2. This stage caps the output voltage to about 14.2 volts. To filter out ripple from the output, the capacitor C8 was also added. This helps prevent the hum signal from being noticed on radio or audio devices.

In constructing the circuit, attach the ICs to a common heatsink close to the pcb. The transistor must be attached to a separate heatsink. The two ICs have built-in protection against short circuits and thermal overload. The TDA2002 can be used in place of TDA2003. The TDA2003 however, has better characteristics. To get a much higher current output from the circuit, the capacitors C4, C5 and C6 must be increased to 2200 μF.

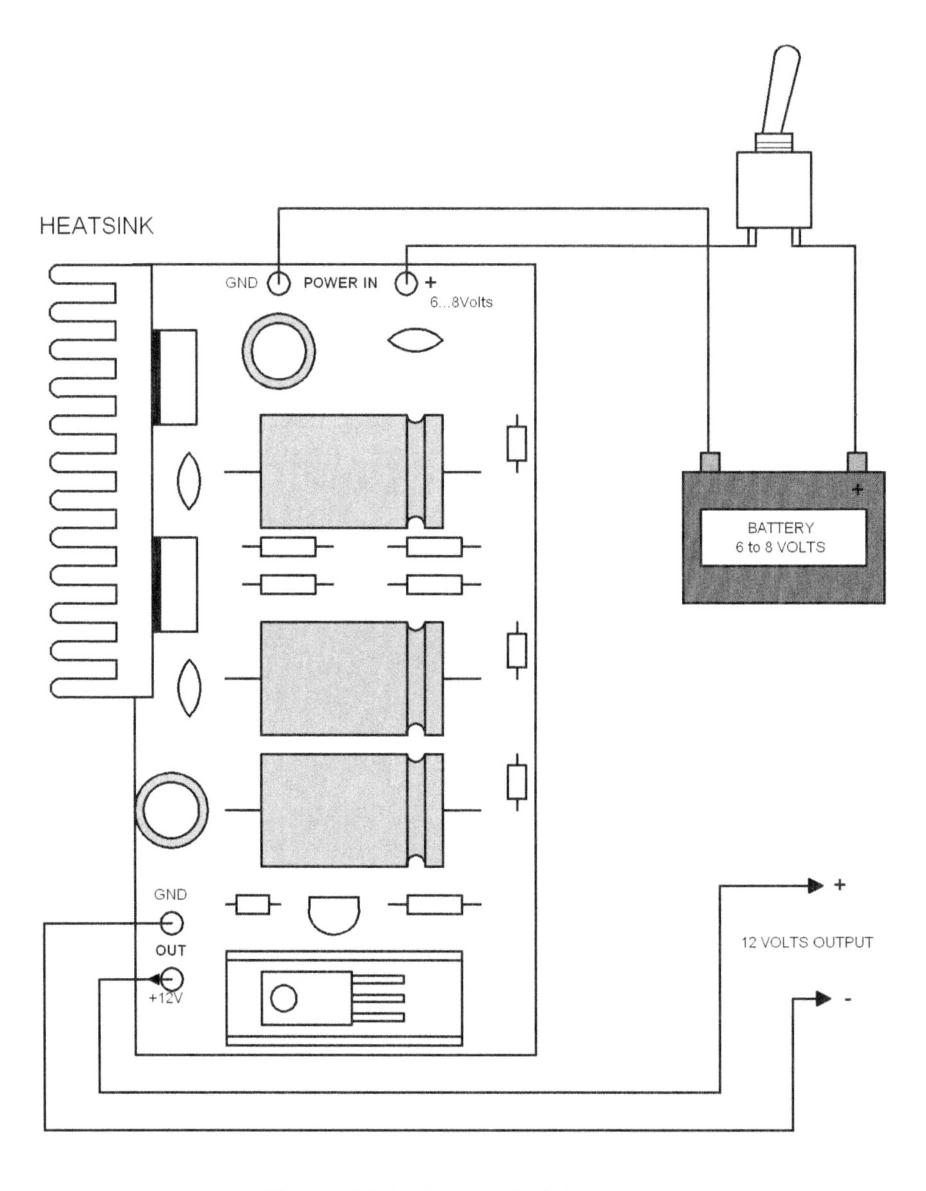

Figure 65.2 External Wiring Layout

66 UNIVERSAL TRIAC CONTROL

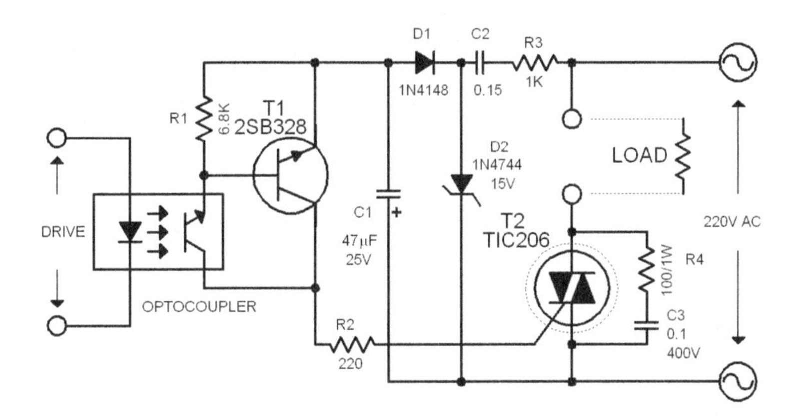

Diagram 66.0 Universal Triac Control

Temperature dependency of electronic components is a well known disadvantage. In the case of triacs, these components are normally less dependent on temperature. However, when the environmental temperature is too low, triacs behave erratically. Actually, the problem lies not much on the triac itself. It simply needs higher gate current at low temperatures. If the triac is driven by an optocoupler circuit, the low temperature affects the optocoupler so badly that it cannot deliver the needed current to the triac. This problem is solved in the featured circuit by adding a transistor at the output of the optocoupler circuit. The transistor amplifies the trigger pulse coming from the optocoupler. The gate current is high enough to trigger the triac in all (temperature) cases.

The optocoupler-transistor circuit works this way: The base of the transistor is driven by the optocoupler. Capacitor C2 works as capacitive bias resistance to avoid power losses. It also helps avoid DC loading of the supply line. The switch current is limited by R3. Diode D1 works as one-way rectifier while C1 works as ripple filter. Diode D2 stabilize the rectified voltage to 15 volts.

Transistor T1 conducts when the optocoupler sends a pulse to its base. Capacitor C1 discharges through the collector-emitter line. The trigger current is limited by R2 to around 40 mA. The discharge time of C1(and with it the pulse length of the trigger) is less than one millisecond.

The RC circuit R4 and C3 protects the triac from voltage spikes. This is very important in all inductive loads.

67 FUSE MONITOR

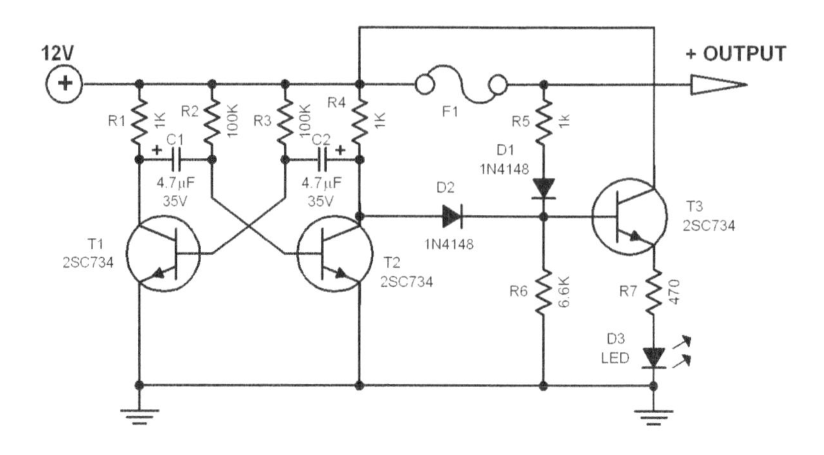

Diagram 67.0 Fuse Monitor

This circuit monitors a DC fuse. Its LED lights continuously when the fuse is intact but blinks if the fuse is busted. The featured circuit is designed for 12 volts but can be modified for other voltages. To use the circuit for 6 volts, divide all resistance values by two. For 24 volts, double all resistance values.

The circuit is made of an astable multivibrator (AMV) and a LED driver. The biggest part of the circuit is before the fuse with the exception of R5. The AMV functions all the time as long as the power is on. The output of AMV is connected to the driver circuit via D2. As long as the fuse is intact, current flows to the base of T3 via R3 and D1 and the LED lights. Once the fuse gets blown, no more current can flow to the base of T3. In this case, the AMV takes over the control of the LED driver and the LED blinks.

The circuit consumes around 25 mA. Most of the current is consumed by the LED. If one decides to use the circuit in battery operated modules, it is highly recommmended to use a high efficiency LED and increase the value of R7 accordingly.

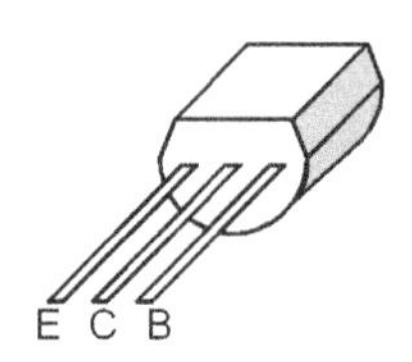

2SC734

68 Automatic Emergency Lamp

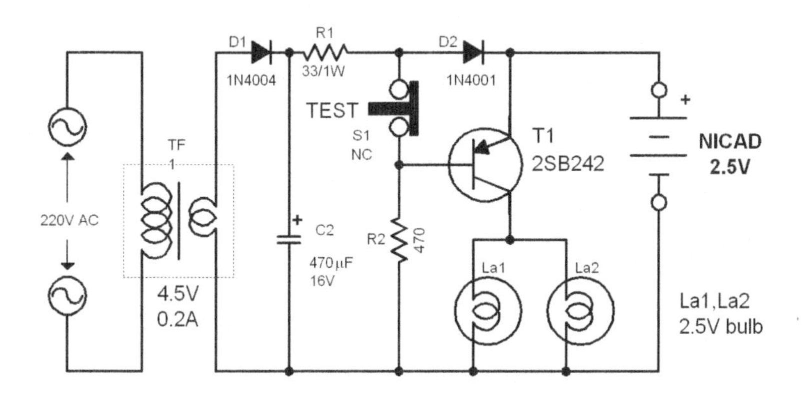

Diagram 68.0 Automatic Emergency Lamp

This circuit automatically turns on a lamp during brownouts. Its power source is a NiCad battery that is being charged continuously by the main power line when there is no brownout. The diagram shows the simplicity of the circuit.

The electronic circuit works this way: the voltage from the step down transformer is rectified by diode D1 and filtered by C2. The NiCad battery gets charged with about 100 mA via the diode D2 and resistor R1. The NiCad battery must have a capacity of at least 2 Ah to tolerate the charging rate. In a normal situation, the base voltage of T1 is positive in relation to its emitter due to the voltage drop at diode D2. Transistor T1 in this case does not conduct and the lamps stay unlit.

When the main power supply fails during a brownout, the charging current is interrupted. In this case, a current flows from the base of T1 via resistor R2 which triggers the transistor to conduct and the two lamps light up. When the main power returns, the charging current flows again through D2 and the transistor turns the lamps off.

The „TEST“ button S1 is used to check the function of the circuit. If the secondary coil of the transformer delivers a higher voltage level, replace the R1 with a higher value resistor to avoid exceeding the maximum charging current allowed for the NiCad battery being used.

69 CENTRONICS PRINTER RESET

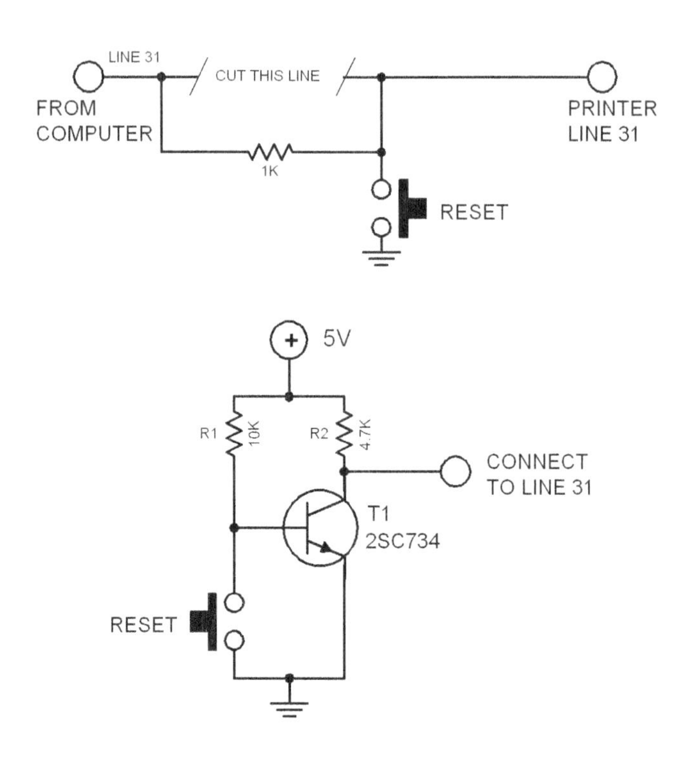

Diagram 69.0 Centronics Printer Reset

This simple circuit is used to reset a centronics printer of older computer systems. The problem is quite well known: once the printer starts to print, it prints up to the last byte that was fed to it. There is no means to stop it from printing except from turning off its power supply, an admittedly unelegant way. However, it is also possible to reset (or stop the printer from printing) the centronics printer by momentarily shorting its pin 31 at the interface connector. The featured electronic circuit shows how to do it safely without causing damage to the computer or the printer.

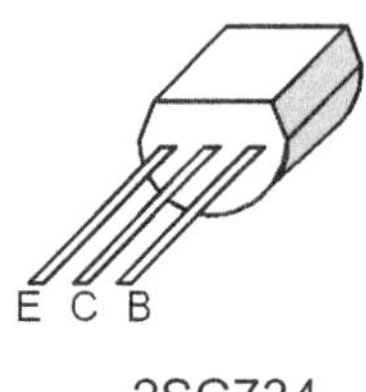

2SC734

70 Multi-Channel A/D Converter

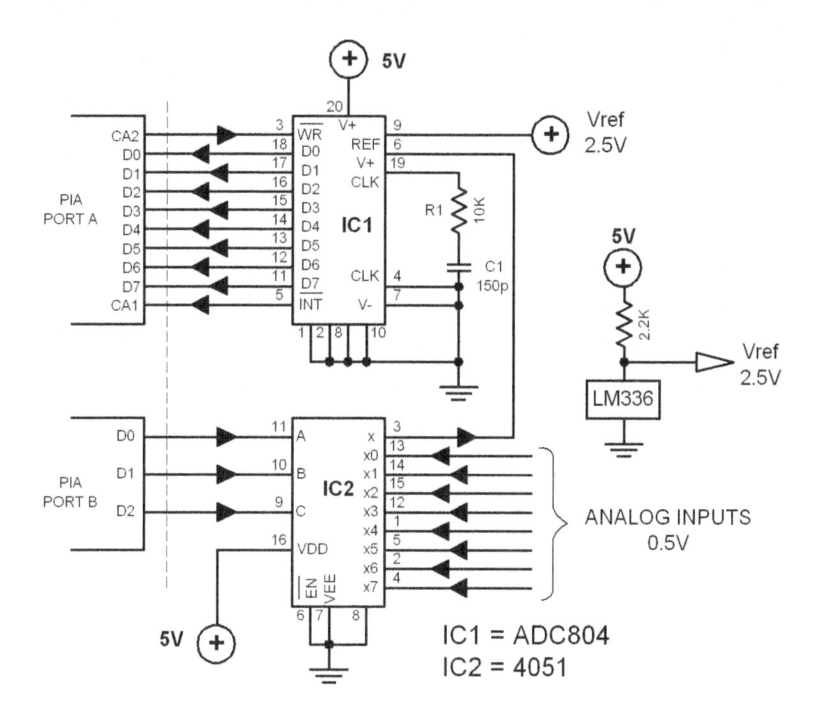

Diagram 70.0 Multi-Channel A/D Converter

If one uses a microprocessor system as a control device, it is almost always unavoidable to monitor several analog signals at the same time. If a high level of accuracy is needed from these monitored signals, a high quality A/D converter is indispensable. It can be very expensive. To solve this cost problem, this featured circuit functions as a multiplexer that switches eight analog inputs.

An analog channel is selected through the inputs A,B and C which are connected to the microprocessor port. At binary 000, channel 0 is selected. At binary 001, channel 1 is selected. And so on. The analog input signal must be between 0 and +5 volts. The selected analog signal is relayed to the pin 3 of the IC2 (CD4051).

The actual A/D conversion is done by CMOS-IC IC1 (ADC804). This IC converts the analog signal into an 8-bit data within 100 µS. The internal clock of the IC needs an RC network at pins 4 and 19. A reference voltage that is exactly at the middle of the measurement range must be fed to pin 9. The output pin 5 (INT) rises up to 5 volts by a falling edge at pin 3. By the next falling edge at pin 3, the conversion begins. 100 µS later, the conversion ends, pin 5 goes down to 0 volts and the 8-bit data can be read from IC1.

Due to the 256 possible combinations, the measurement resolution within a 5 volts range is 19 millivolts.

This electronic circuit can be connected to any microprocessor system with port modules. The programming is of course dependent on the particular application. One can practically use Basic commands such as POKE and PEEK in applications that are relatively slow like for example a heater control or alarm systems. For faster applications, it is necessary to write the program in machine language.

71 X/Y PLOTTER INTERFACE

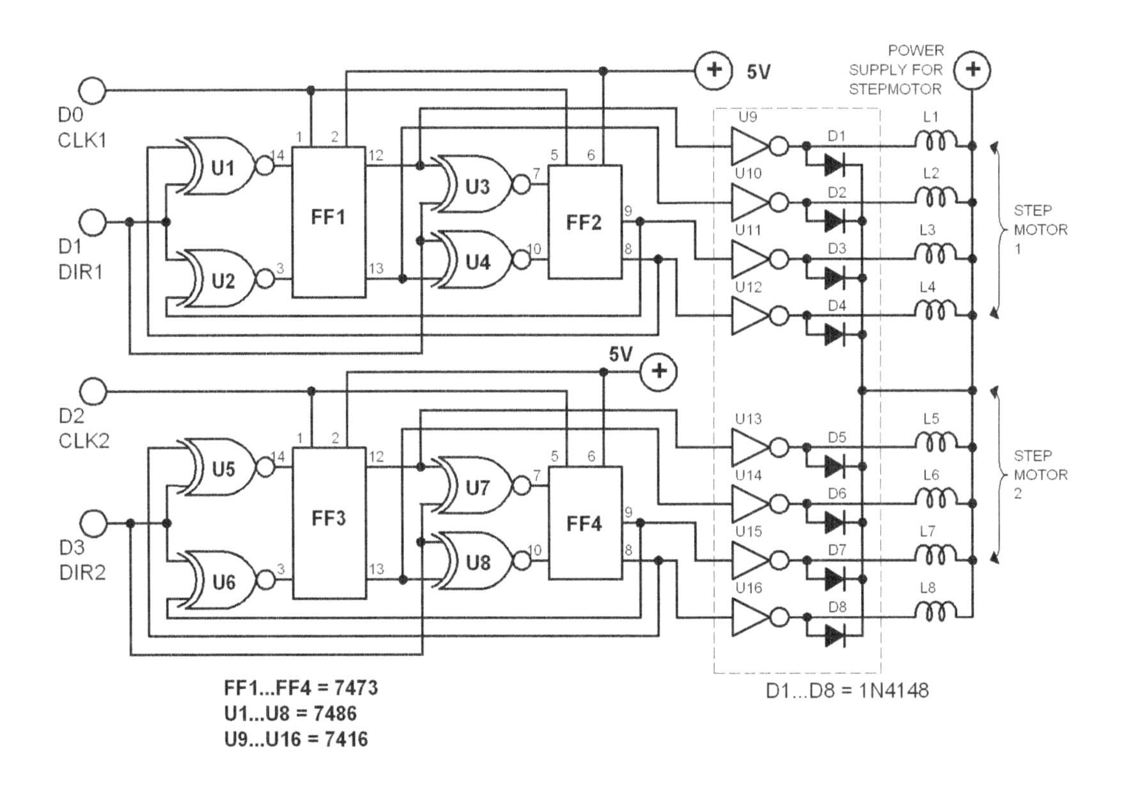

Diagram 71.0 X/Y Plottter Interface

This circuit can drive two step motors and a relay based on parallel data. At the same time, the circuit relays the position of two microswitches that are attached to the step motors. This way, it sends information about the status of the step motors back to the computer. This makes the circuit highly practical for robotic projects or x/y plotters- printers.

The circuit has two S/R gates functioning as memory to store the signal from the microswitches S1 and S2. This information is read by the computer through D5 and D6 data lines. The relay and its associated driver circuitry enables control of a pen or any other device through line D4. Control of the two step motors is done via four XOR gates U1 to U8. The direction of rotation is controlled via D1 and D3 lines.

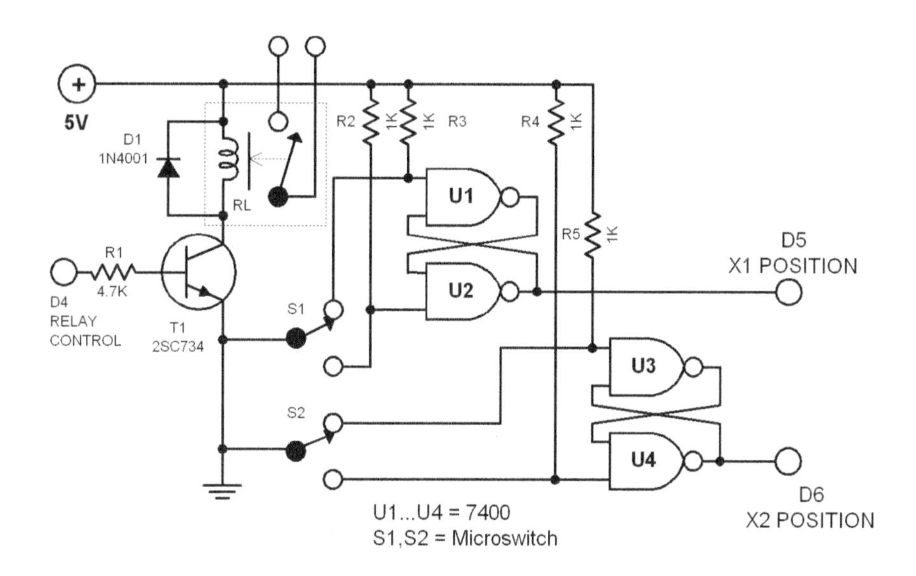

Diagram 71.1 X/Y Plottter Interface

The clock signal is fed to the driver circuit via D0 and D2 lines. The ULN2803 IC (U9-16) delivers a maximum current of 500 mA. The maximum motor voltage allowed is 50 volts. As shown in the diagram, protective diodes are integrated inside the ULN IC.

The interface circuit needs two power supplies: 5 volts for the interface and 12 volts for the step motors. Motor 1 is shown in the diagram as L1 to L4 and motor 2 is shown as L5 to L8. In programming the computer to drive the circuit, set the I/O port to 5 output bits (D0 to D4) and 2 input bits (D5 and D6) and send the appropriate data signals to the driver interface.

72 220V POWER INTERFACE

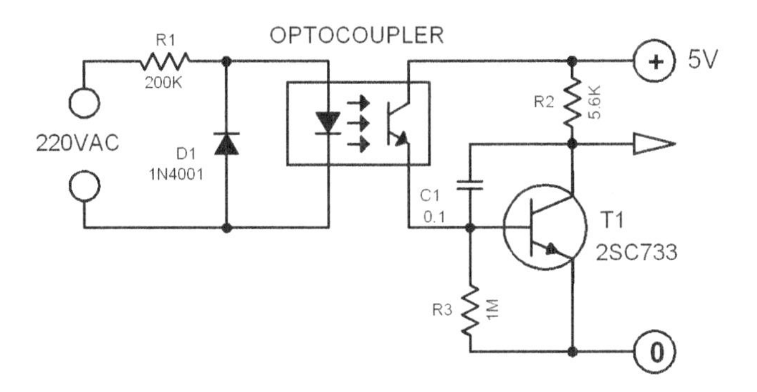

Diagram 72.0 220V Power Interface

This simple electronic circuit is intended as an interface for monitoring electric equipments and devices using a computer. The interface only senses whether the device being monitored is turned „on" or „off". The most important aspect of the circuit is the galvanic isolation between the AC main line being monitored and the interface to the computer. This is done with the use of the optocoupler IC. The original optocoupler used was a TIL111 but a suitable replacement can be used too.

The circuit is very simple. To avoid having to modify the circuit for each device being monitored, the circuit monitors the device's AC power line directly. The resistor R1 lets a current of around 0.5 mA through the optopcoupler LED. The other half of the current is rectified through the diode D1. It is obvious that the current to the optocoupler LED is half wave rectified. This means that the phototransistor part of the optocoupler (the receiver part) receives only light impulses of around 100 µA. This is however strong enough to drive the transistor T1. The capacitor C1 filters out the current pulsation and maintains a smooth output current. In case the current pulsation is needed, just remove the capacitor C1 from the circuit.

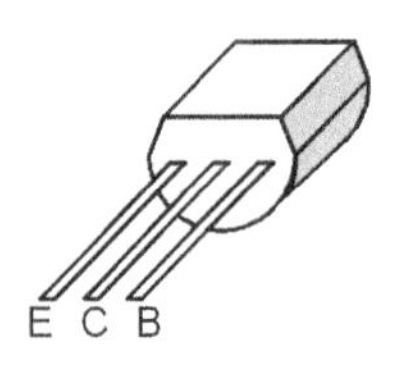

2SC733

73 SYMMETRICAL POWER SUPPLY

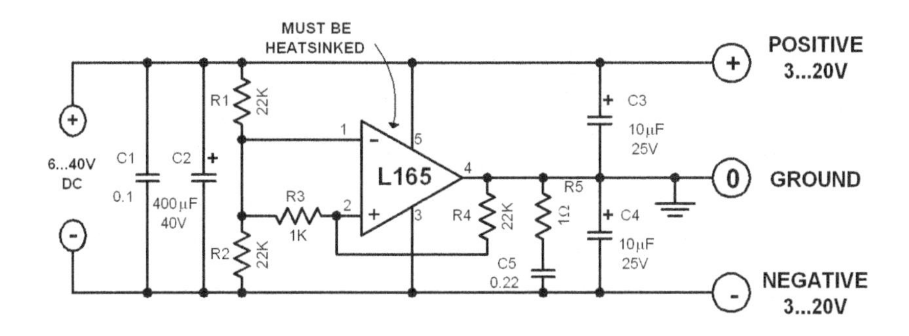

Diagram 73.0 Symmetrical Power Supply

The compact 5 pin L165 IC generates a stabilized symmetrical power supply from a single asymmetrical power supply. The output voltage is however, half of the input voltage. One only needs to add the ripple filter capacitors C1,C2, C3 and C4 and some resistors for setting the symmetry. In constructing the circuit, place the capacitors C1 and C2 as close to the IC as possible. On the other hand, place the capacitors C3 and C4 close to the output jacks. Make sure that the circuit lines on the pcb are properly dimensioned to handle high current levels. Current levels up to 3 amperes can flow through the circuit. Additionally, provide a proper heatsink for the L165 IC.

The IC can also be viewed as a voltage amplifier. It amplifies the voltage appearing at the junction between R1 and R2.

One can also replace the IC with TCA1365. However, when using the TCA IC, pins 3 and 4 have to connected together. Also, connect a 220 pF capacitor between pins 5 and 6.

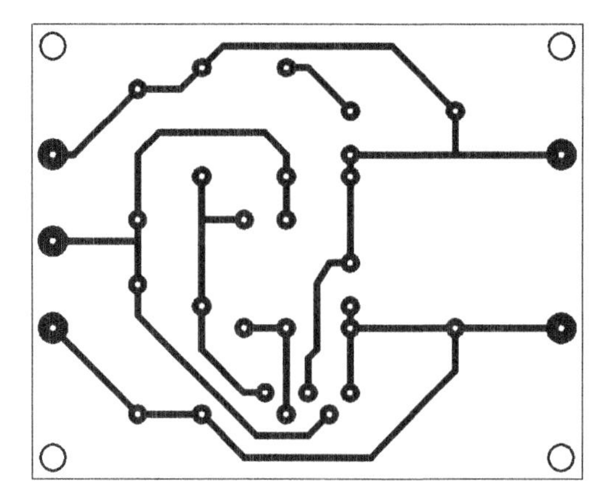

Figure 73.0 Printed Circuit Layout

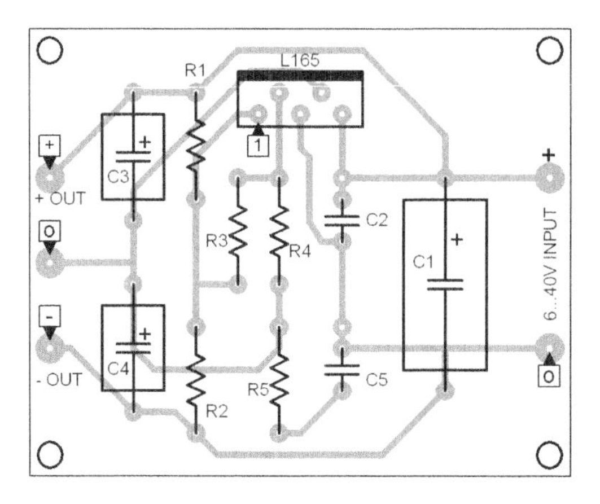

Figure 73.1 Parts Placement Layout

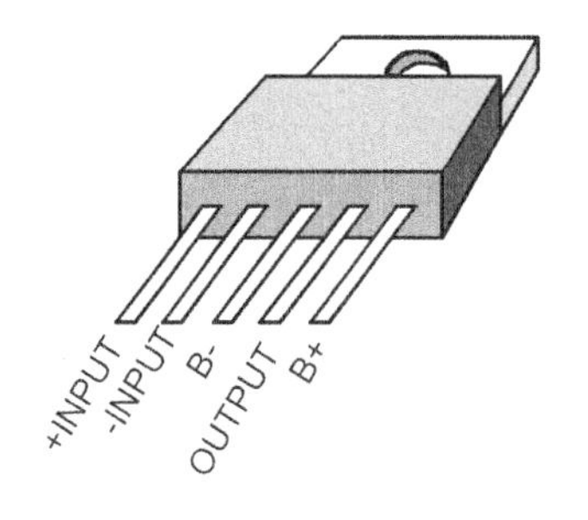

L165

Pin configuration

1 = + IN
2 = - IN
3 = B- (- VS case)
4 = OUT
5 = B+ (+ VS)

74 KEYBOARD TESTER

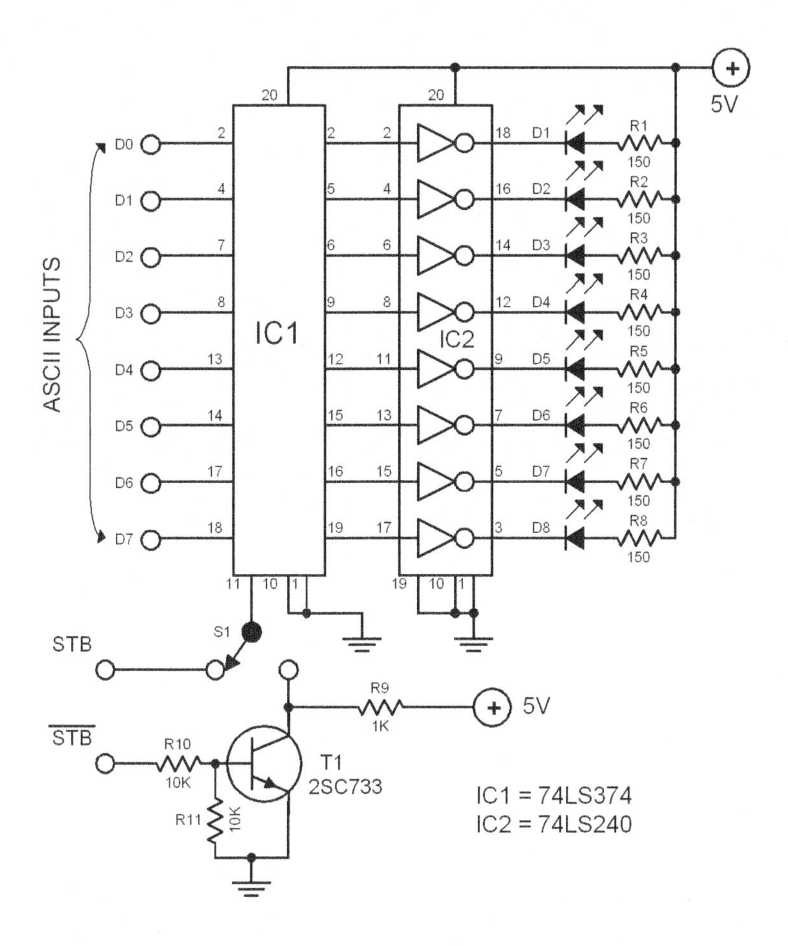

Diagram 74.0 Keyboard Tester

This circuit is designed to test keyboards with an ASCII bus connection to the computer. To determine the pin designations of such an ASCII keyboard, it is practical to know first which pin is the power supply line. This is usually easy to do. The keyboard case is opened and the flat cables are followed until the power supply line is found. Next, the strobe pin must be found. After connecting the power supply, find the strobe line by using a logic probe, a voltmeter or an oscilloscope. The line that shows a short pulse every time a key is hit, is the strobe line. If an oscilloscope is being used for this test, it is easy to find out whether the strobe pulse is negative or positive going.

Once the power and the strobe lines are found, the keyboard tester can do its work. Attach all the data output lines (all lines except the power and strobe) from the keyboard to the keyboard tester circuit. Switch S1 is used to select between negative and positive going strobe pulse. A positive going strobe is fed directly to the IC1 at pin 11 (CLK). A negative going strobe is fed first to the inverter circuit composed of T1, R9, R10 and R11 before it reaches the pin 11 of IC1.

The strobe pulse triggers the D flip flops of IC1 every time a key is hit. This action causes the data appearing on the output lines to be stored by the flip flops. The flip flop outputs are also set accordingly. The following IC2 holds eight inverting drivers. If one of the flip flops has a high level, the driver's output is low. This causes current to flow through the connected LED making it light up.

To determine the correct order of the data output lines, an ASCII table is needed. One hits a known key and notes down the lighted LEDs. The pattern is then lookep up at the ASCII table and the correct order of the output lines can be found.

75 SYNC SIGNAL SEPARATOR

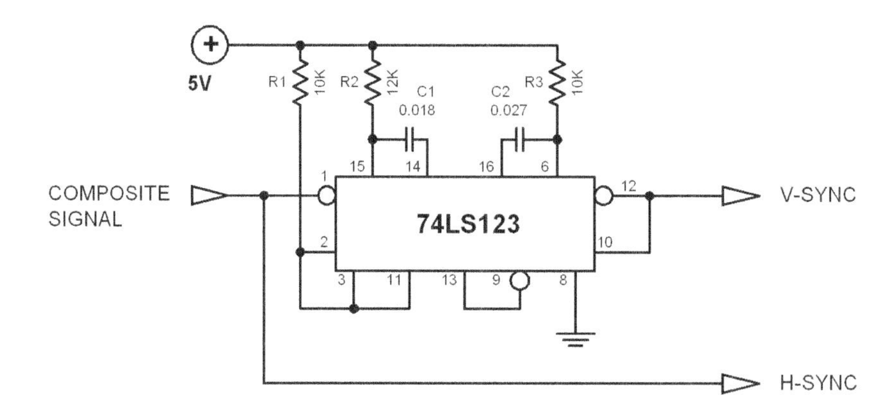

Diagram 75.0 Sync Signal Separator

Many monitors have separate inputs for vertical and horizontal sync signal. This becomes a problem when the computer has only a single line for a combined sync signal. The solution to that problem is this featured circuit. This simple circuit separates the composite sync signal into two separate horizontal and vertical sync signals.

The composite signal can be directly used to drive the monitor's horizontal sync input. That is why the composite signal line is connected directly to the H-sync line. However, to get the vertical sync signal, the double monoflop 74LS123 is needed as shown in the circuit diagram. The first monoflop time is somewhat longer than the distance between two line scan pulses. All horizontal line sync pulses are supressed. The second monoflop delivers only vertical sync pulses.

76 PULSE GENERATOR

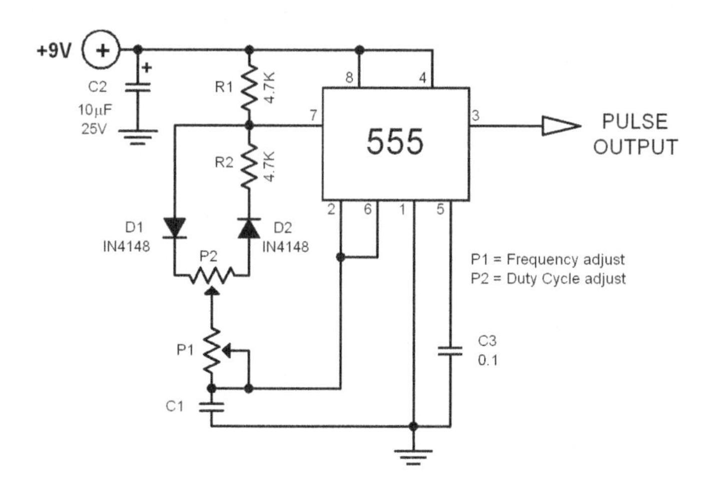

Diagram 76.0 Pulse Generator

This is a pulse generator with adjustable duty cycle made with the versatile 555 timer IC. The circuit is basically an astable multivibrator with a 50% pulse duty cycle. One thing that differs the circuit from the standard design of a 555 timer is the resistance between pins 6 and 7 of the IC composed of P1,P2,R2,D1 and D2.

The diodes D1 and D2 set a definite charging time for capacitor C1 which produces a 50% duty cycle in a normal case. The duty cycle (n) is dependent on P1 and P2 in the following manner:

$$n = 1 + P2/P1$$

An example: If P2 = 0 (n = 100%) then the frequency can be approximately calculated with the following formula:

$$f = 0.69 /((2*P1 + P2 + 4.7\text{Kohm})*C1)$$

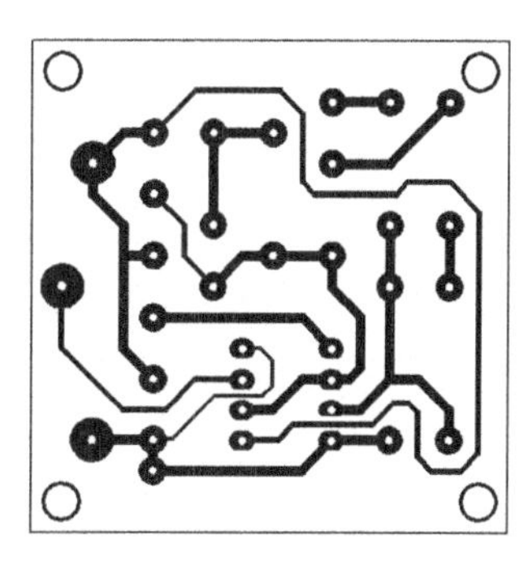

Figure 76.0 Printed Circuit Layout

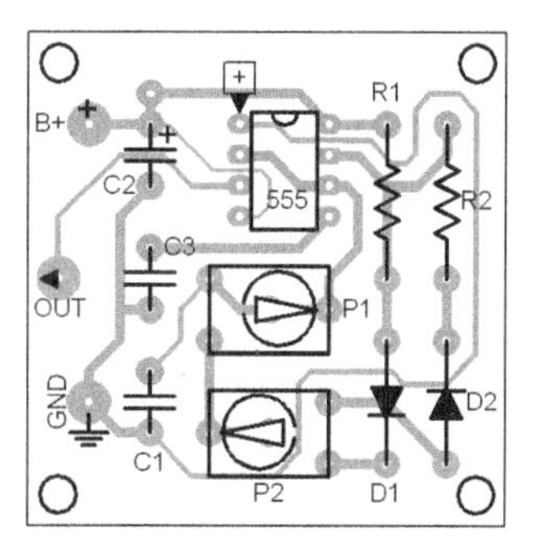

Figure 76.1 Parts Placement Layout

77 SQUAREWAVE GENERATOR

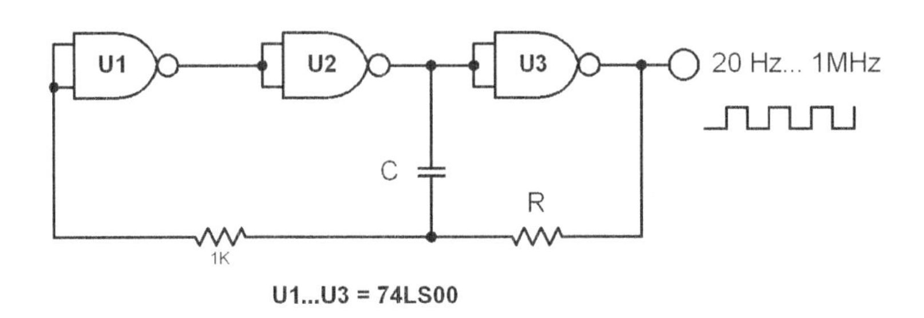

Diagram 77.0 Square Wave Generator

You can construct a square wave generator in a simple way by using three TTL gates. This generator can be used in many different applications. The generator oscillates within a wide frequency range. Its stability is good enough for most applications. It starts smoothly, its construction is simple and it is highly independent from the voltage supply.

The output frequency is dimensioned through the RC components and time delay of the three inverter gates (in this case, NAND gates with their inputs shorted together). The time delay of a logic gate is the elapsed time between a change of input state and the resulting change of the output state. This time is highly dependent on the temperature and the voltage supply. Therefore, the time delay must be prevented to affect the oscillation frequency. In every period of the oscillation signal, the ouput of the gate switches twice (from „1“ to „0“ and back). The total time delay of the three inverter gates in series multiplied by 2 in the formula. The oscillator frequency is conventionally labeled *fo*. To prevent the temperature and voltage supply from affecting the frequency, this fo must be smaller than 1/2tpn where:

tp = average delay time of each gate
n= the number of gates.

The above described oscillator has a tp of 10ns and n = 3. The frequency is therefore:
fo << 1/2tpn = 1/ 2.10ns.3 = 16.6 MHz.

The voltage at the gate inputs varies from about +6 volts to -4 volts. The oscillation frequency can be made variable by using a 2.2 potentiometer for R.

78 MULTIPLEXER SWITCH

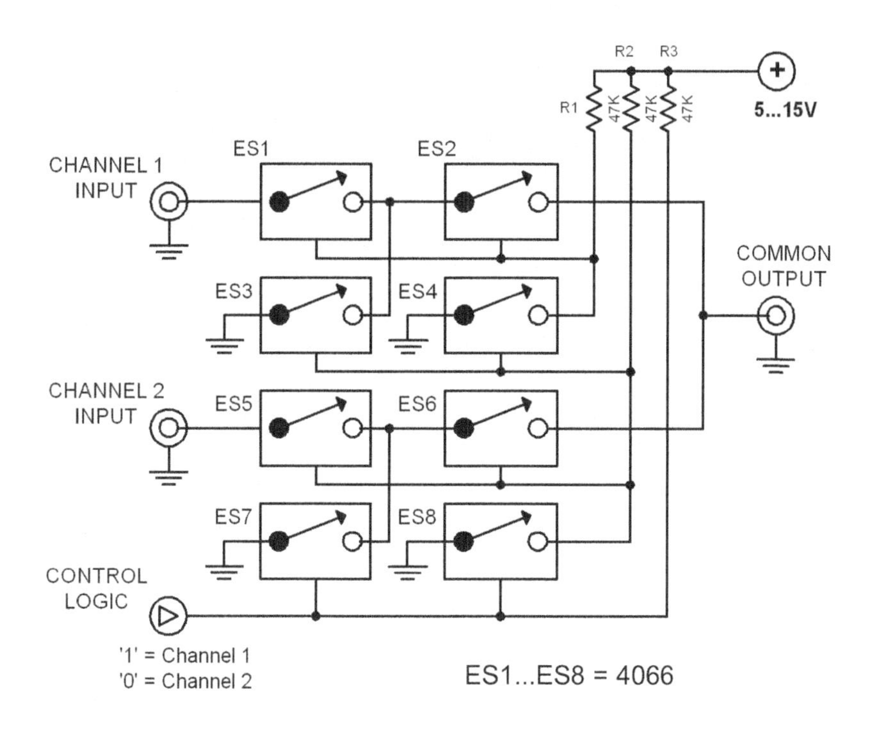

Diagram 78.0 Multiplexer Switch

This circuit was designed to switch multiple signal sources into a single common output. It is a replacement for the conventional mechanical switches. Conventional multiplexer switches have an annoying disadvantage that the video signals mix over when the source is changed. The featured circuit avoids this problem by disabling the inactive signal line.

The electronic circuit is simple. The heart is a digitally controlled analog switch IC (4066). The switches are labeled ES1 ... ES8 in the diagram. Each 4066 IC contains four switches. When channel 1 is selected, switches ES1 and ES2 are closed and ES3 is open. Channel is suppressed at this moment since ES5 and ES6 are open and ES7 prevents a cross-over between the two channels by grounding the channel 2 line. Since one IC is used for each channel, the cross-talk between the two channels is further prohibited.

The bandwidth of the circuit is around 8 MHz. The current consumption is around 1 mA (dependent on the power suppy voltage).

A high power supply voltage level is recommended since the input impedance level of the circuit drops with higher voltage level.

When the multiplexer circuit is connected to a 75 ohms impedance load, some signal losses will result due to the internal resistance of the switches. This can be compensated by connecting a video amplifier at the output of the multiplexer circuit.

79 LED DIMMER

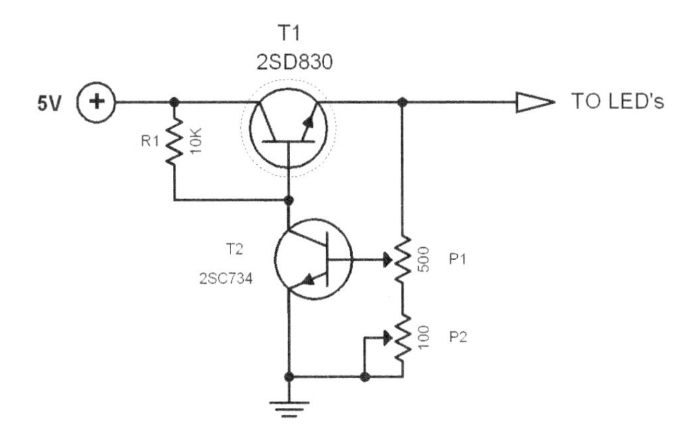

Diagram 79.0 LED Dimmer

Typical segment display LEDs consume around 25 mA for each segment. In fact, each segment must be limited to a maximum of 25 mA current. This current limit is traditionally done by adding resistors in series to the segments. If a six digit display is to be current limited through this traditional way, at least 42 series resistors are needed. This adds additional complication to the pcb design not to mention the extra space needed on the board.

Another disadvantage is the lack of capability to adjust the brightness of the segment display once the resistors are in place. To counter the above described problems, the feature circuit was designed. It allows for eventual adjustment of the display's brightness. The series resistors are not needed anymore for the segment display which simplifies the pcb design and construction. The circuit is simply a voltage regulator with variable voltage output. The segment display LEDs are connected to its output. The brightness of the LEDs are dependent on the output voltage. Since the voltage regulator is variable, the brightness of the LEDs is also variable. Potentiometer P1 is used for rough adjustment while P2 is used to trim the brightness in finer resolution. The output of the regulator circuit can be varied from 0 to 4.3 volts.

Before turning on the circuit, set the potentiometers to zero point. Then after the circuit is turned on, slowly adjust the potentiometers until the desired brightness is achieved. In a typical six segment display, the maximum current from the regulator circuit must not exceed 1 ampere. One must be careful due to the fact that this value can be easily exceeded. For example: if one segment consumes 25 mA each, then 7 segments of a six digit display will consume around 1050 mA. That is a bit more than 1 ampere! The transistor T1 must be heatsinked because high current consumption will produce a lot of dissipated heat.

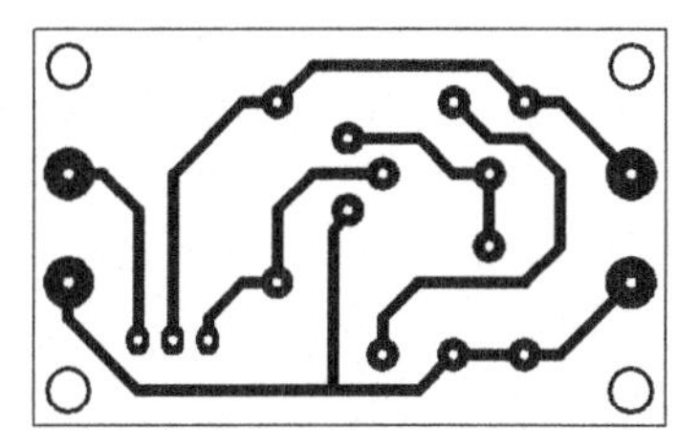

Figure 79.0 Printed Circuit Layout for the LED Dimmer

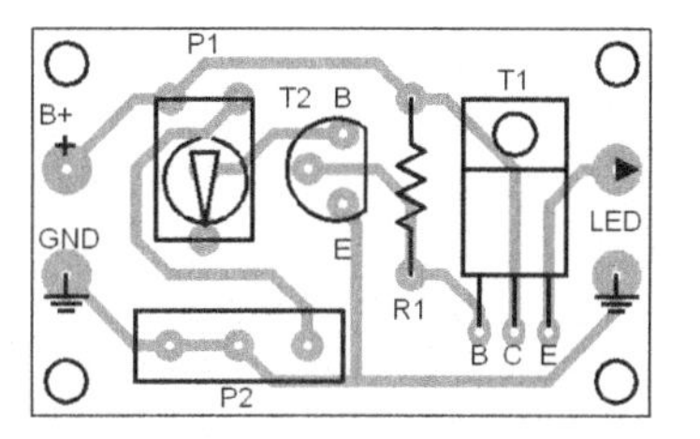

Figure 79.1 Parts Placement Layout for the LED Dimmer

80 MUSICAL DOORBELL

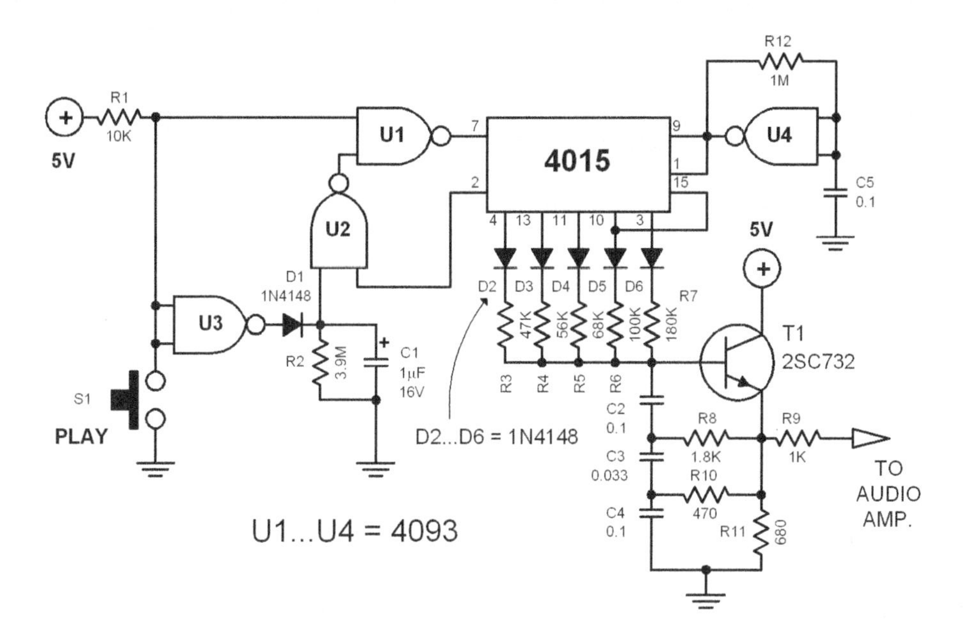

Diagram 80.0 Musical Doorbell

In many cases, a doorbell that sounds off a musical tone is preferable over the common buzz sound. This featured circuit is a musical doorbell. After the button S1 is pressed, a short melody is played. When the button is pressed many times in rapid succession or pressed longer, a different melody is generated and the melody plays longer.

The circuit works this way: when the button S1 is activated the inputs of U3 and one input of U1 switches to logic „0“. The data input (pin 7 of IC 4015) becomes logic „1“. The 4015 is a static 4-bit shift register. Each clock impulse coming from U4 shifts this logic „1“ further in the register. The clock frequency is around 5 Hz.

The number of shifted logic „1" is directly dependent on the length of time the switch S1 is closed. Once at least one shift register is logic „1", a current flows to the base of T1 through a corresponding resistor. The transistor T1 functions as a current controlled oscillator. The tone pitch is dependent on the logic state of the different flip-flop outputs of the shift register. Each clock pulse shifts the logic „1" in the register by one flip-flop. When S1 is pressed one more time at this moment, another logic „1" is added to the register. One output of the register (pin 2) is coupled back to U2 and U3 so that all the logic „1" in the register always run in a loop.

When S1 is released (opened), the register runs until the capacitor C1 gets discharged through R2. When S1 is again pressed (closed), the capacitor C1 stays charged causing the musical bell to play continuously.

The difference between the two ways of activating the switch S1 is that different combinations of logic „1" are inputted into the shift register. These different combinations produce the different melodies played by the circuit. The musical doorbell must be connected to an audio amplifier. The supply voltage is not critical. It can be between 5 and 15 volts. The circuit consumes around 15 milliamperes.

81 FUNCTION GENERATOR

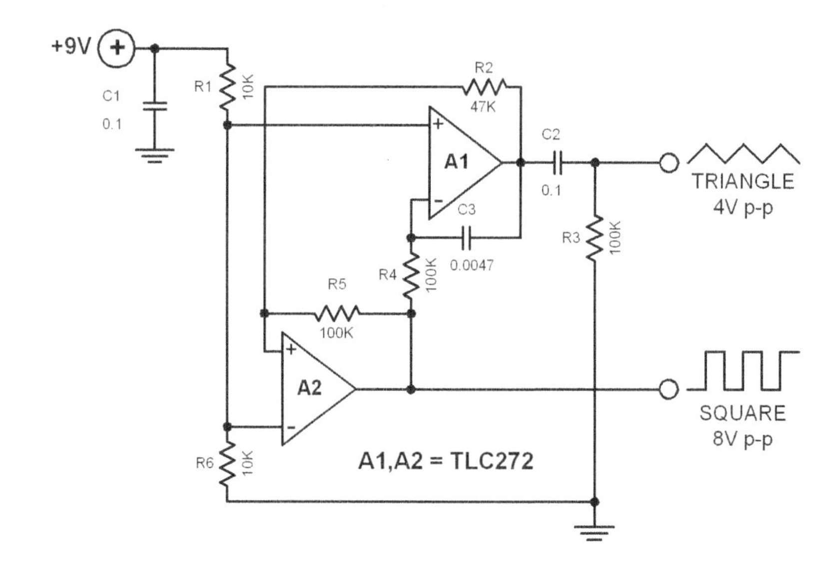

Diagram 81.0 Function Generator

You can build a function generator in a very easy way by using this circuit. It is simple to construct. It generates a triangle wave and a square wave. It is powered by a single voltage supply. It can also be made variable.

The heart of the circuit is the TLC272 that is manufactured by Texas Instruments. It is a modern CMOS opamp that consumes very little current. Principally, the circuit is made of two opamp circuits. The A1 part is an opamp configured to function as a schmitt trigger. It triggers at around 4.5 volts reference voltage. The opamp A2 is configured as an integrator. It converts the triangle wave output of A1 into a square wave signal.

The oscillation frequency of the circuit is dependent on the RC components. One can also replace the fixed value resistor R4 with a potentiometer to vary the signal frequency. The following formula: fo = 1/4RC x R2/R1 will give the oscillation frequency.

The circuit must be connected to a high ohmic load, at least 10K. The function generator can be used for the entire audio frequency band.

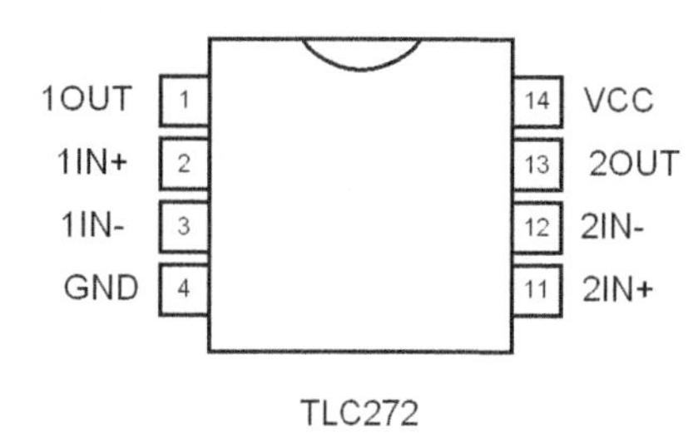

Figure 81.0 TLC272 Pin Package (Top view)

Most simple function generators generate sinewave, squarewave and trianglewave but a clean sawtooth is rarely available. This circuit forms a sawtooth by combining a squarewave and a trianglewave signal together. The quality of the sawtooth is therefore dependent on several factors: the linearity of the trianglewave, the edges of the squarewave and the phase relation between the trianglewave and the squarewave.

The converter is constructed out of a single opamp which mixes the two input signals. The trianglewave is directly fed to the inverting (minus) input of the opamp while the squarewave is first processed by the FET T1. Out of the opamp output comes a trianglewave with an inverted falling edge. The available sawtooth signal has a doubled frequency. When the DC level of every inverted edge is raised to a certain value so that its lowest level synchronizes with the highest level of the previous edge, a sawtooth with the correct frequency and doubled amplitude can be generated. This quite complicated technique can be easily applied by adding the output signal with the squarewave signal. P1 and P2 are responsible for mixing the two signals. Resistors R2 and R4 are 1% types.

The converter delivers a clean sawtooth signal between 60 Hz and 15 kHz. The supply voltage can be between +/-10 volts and +/-15 volts. The current consumption of each opamp is around 4 to 6 mA.

82 BRIGHTNESS REGULATOR

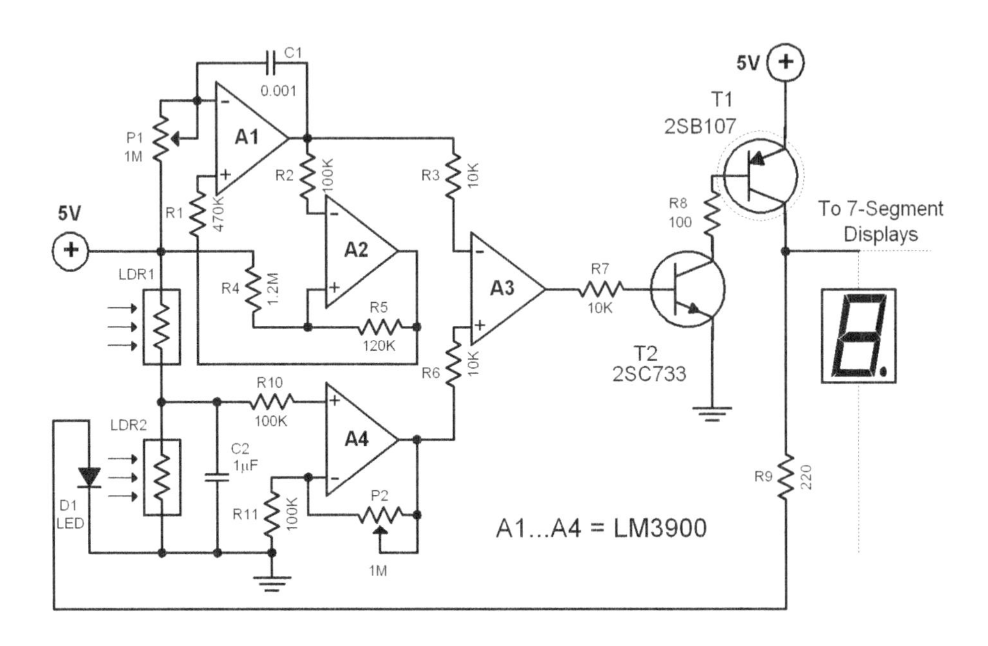

Diagram 82.0 Brightness Regulator

This dimmer circuit automatically adapts the brightness of a 7-segment LED display to the brightness of the surrounding light. Also, the supply voltage of the LED display is rapidly switched on and off by a power transistor. The switching frequency is around 1 kHz, fast enough to fool the human eye that the LED is continuously lighted. The brightness of the LED display is dependent on the ratio of the „on“ period to the „off“ period of its voltage supply. For example, when the transistor T1 is „on“ for the same time period as it is „off“ (50% on and 50% off), then the average voltage at the LED display will be equal to one half of the supply voltage and the LED display will light with somewhat reduced brightness.

The circuit compares the brightness level of the surrounding light with that of the brightness level of the LED display. The brightness level of the LED display is sampled through the D1/LDR2 components which are connected to the supply line of the dispay. This arrangement is needed since it would be impractical to place the LDR2 in front of the LED display. The other LDR2 samples the surrounding light.

The opamps A1 and A2 work together as a sawtooth generator. Its frequency can be set through P1. The voltage from the LDR junction is amplified by opamp A4. Opamp A3 compares the sawtooth voltage with the output voltage of A4.

This dimmer circuit automatically adapts the brightness of a 7-segment LED display to the brightness of the surrounding light. Also, the supply voltage of the LED display is rapidly switched on and off by a power transistor. The switching frequency is around 1 kHz, fast enough to fool the human eye that the LED is constinuously lighted. The brightness of the LED display is dependent on the ratio of the „on" period to its „off" period of its voltage supply. For example, when the transistor T1 is „on" for the same time period as it is „off" (50% on and 50% off), then the average voltage at the LED display will be equal to one half of the supply voltage and the LED display will light with somewhat reduced brightness.

The circuit compares the brightness level of the surrounding light with that of the brightness level of the LED display. The brightness level of the LED display is sampled through the D1/LDR2 components which are connected to the supply line of the dispay. This arrangement is needed since it would be impractical to place the LDR2 in front of the LED display. The other LDR2 samples the surrounding light.

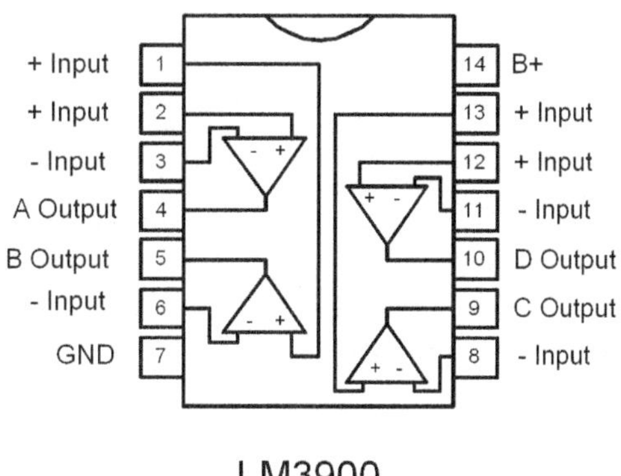

LM3900
QUAD OPAMP

The opamps A1 and A2 work together as a sawtooth generator. Its frequency can be set through P1. The voltage from the LDR junction is amplified by opamp A4. Opamp A3 compares the sawtooth voltage with the output voltage of A4. Depending on the ratio of the voltage from A4 and the sawtooth voltage from A2, the LED display supply voltage is turned on or off via T1 in a fraction of each sawtooth cycle.

The maximum brightness of the LED display can be changed by either adjusting the distance between LDR and D1 or by inserting a mask between them. The switch transistor T1 can deliver a maximum current of 1 ampere. T1 must be heatsinked. If the LED segment display has a „strobe" input, the transistor T1 can be removed and the transistor T2 connected directly to this „strobe" input.

Parts List

R1 = 470K
R2, R10, R11 = 100K
R3, R6, R7 = 10K
R4 = 1.2M
R5 = 120K
R8 = 100Ω
R9 = 220Ω
C1 = 0.001µF/60V
C2 = 1µF/10V
D1 = LED

T1 = 2SB107
T2 = 2SC733

A1...A4 = LM3900

P1, P2 = 1M

LDR1, LDR2 = Light dependent resistor

83 ALARM SIMULATOR

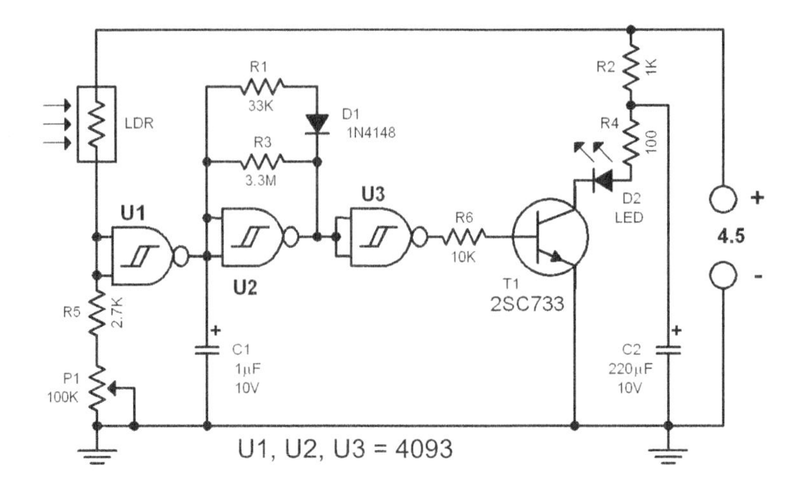

Diagram 83.0 Alarm Simulator

This circuit simulates an alarm to discourage thieves from their criminal intentions. It is originally designed to be used on bicycles but can be used for cars too. It blinks a LED on and off thus simulating a real alarm circuit.

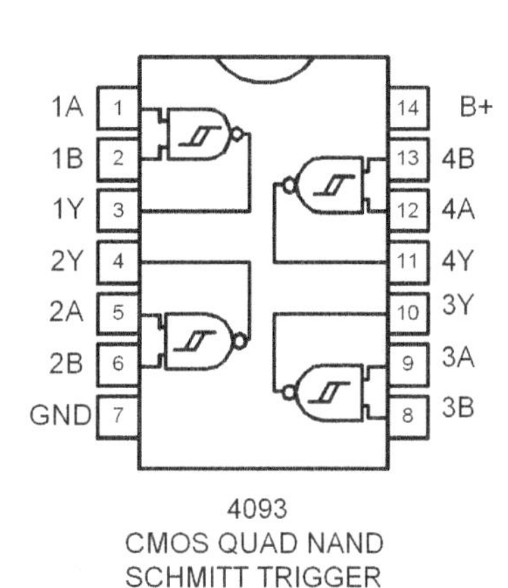

The electronic circuit consists of a square wave oscillator generating a 1 MHz signal with a pulse width of 10 milliseconds. The LED lights up 10 milliseconds long. The circuit consumes 30 milliamperes when the LED lights. The average current consumption is therefore 30mA/100 = 300 microamperes. This value can be decreased furthermore when the circuit is turned off during the day. To switch off the circuit automatically, the circuit has a built in LDR. When the LDR is exposed to light, it becomes low ohmic and the output of U1 becomes 0. The oscillator stops and the LED turns off.

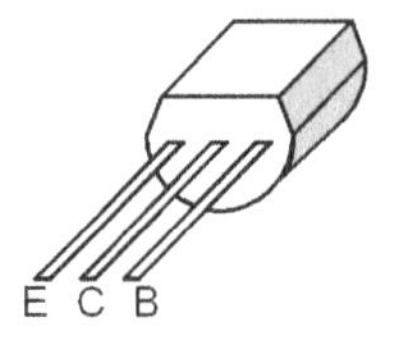

2SC733

The light sensitivity of the circuit is set with the potentiometer P1. When there is no light falling on the LDR, the input of the U1 is connected to the ground and its output becomes „1" thereby starting the oscillator. The oscillator drives the LED to blink. If the circuit is powered with a 4.5 volts battery that has a 2 Ah capacity, it will work for around 1.5 years.

84 TRIGGERED SAWTOOTH

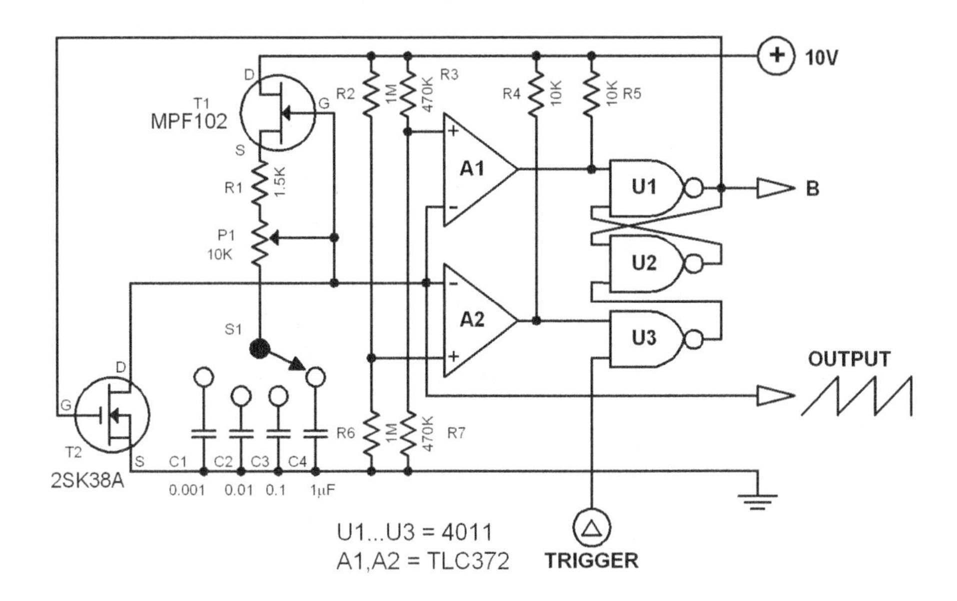

Diagram 84.0 Triggered Sawtooth

This sawtooth generator can be used in many applications. One such application is a horizontal time deflection unit for an oscilloscope. The FET transistor T1 is configured as a current source that charges one of the capacitors C1 to C4 depending on the switch S1 position.The increasing voltage at the capacitor is linear and is compared with two reference voltages (at R3/R7 and R2/R6) with the help of two comparator ICs (A1 and A2). Once the charge voltage is above +5V (comparator level at A1), the flipflop U1/U2 is reset. This results to the capacitor being discharged through transistor T2. When the capacitor voltage sinks below the lower comparator level (A2 level), the output voltage of A2 jumps to +10V. At this moment, it is possible to set the flip flop through the trigger input. The transistor T2 turns off and the charging cycle at the capacitor begins again.

The U1,U2,U3 ICs are designed to guarantee that the trigger impulse can only start the sawtooth signal when the capacitor is really discharged. This technique prevents accidental in-between triggers.

The generator output can only be connected to a high impedance circuit. A low impedance would otherwise distort the waveform. The current consumption of this circuit is around 6 mA.

85 REFRIGERATOR ALARM

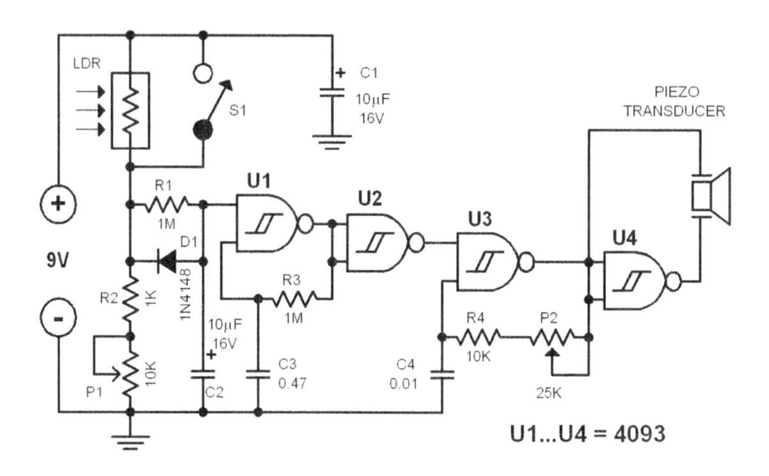

Diagram 85.0 Refrigerator Alarm

It is a widely accepted fact that refrigerators must remain closed so that the cold air cannot escape. However, it happens every now and then that one forgets to close the refrigerator completely.

The featured electronic circuit prevents this from happening by sending out an alarm signal every time the door is left open for some time. The sensor of the alarm is an LDR. After a short delay, it triggers an alarm signal and keeps the signal on until the door is shut close. Another application of the circuit is to monitor other doors. For example, it can monitor house doors in winter to prevent heat loss due to doors left open. However, in this case, the LDR must be replaced with a switch that is attached to the house door. This switch must close when the door is opened.

The refrigerator alarm comes very handy in reminding us to close the fridge correctly.

The alarm circuit functions this way: When a light falls on the LDR, it becomes low ohmic. The voltage at the junction of LDR/R2 charges the capacitor C2 via the R1 slowly. After about 10 seconds, the capacitor voltage reaches the input threshold of U1. Then U1 generates a square wave signal with a frequency of several Hertz. The time delay can be shortened by decreasing the value of R1 down to 220K. U2 inverts the signal from U1. This inverted signal switches the U3 tone oscillator. When there is no light hitting the LDR, U3 is turned off. Finally, U4 amplifies the tone signal generated by U3 and drives the ceramic transducer. The frequency is doubled at the transducer. This is due to it being coupled to the input and outpuf of U4.

The light sensitivity of the LDR can be adjusted with P1. The alarm tone frequency can be adjusted to the resonant frequency of the ceramic transducer through P2.

The maximum volume level can be reached when the tone frequency is equal to the ceramic transducer's resonant frequency.

The circuit can be powered with a small 9-volts battery. The current consumption is about 0.6 mA at standby and increases to about 5 mA when the alarm activates.

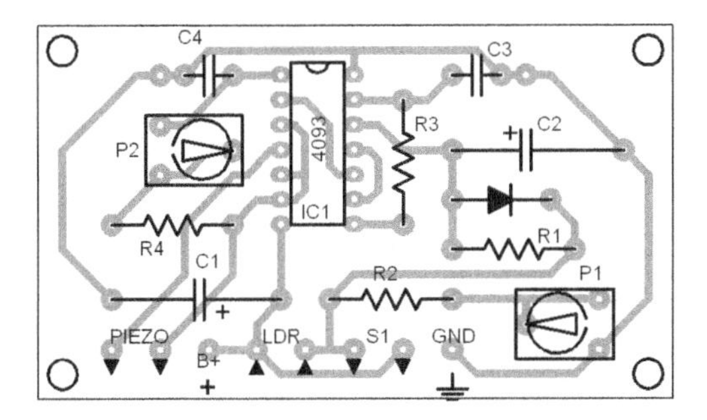

Figure 85.0 Parts Placement Layout

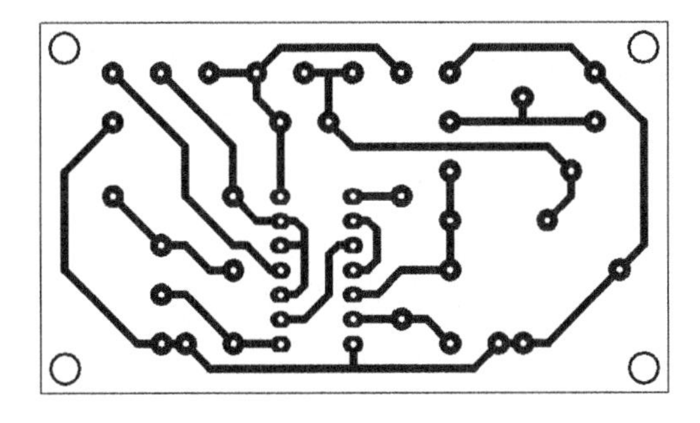

Figure 85.1 Printed Circuit Layout

86 POWER LINE REMOTE CONTROL

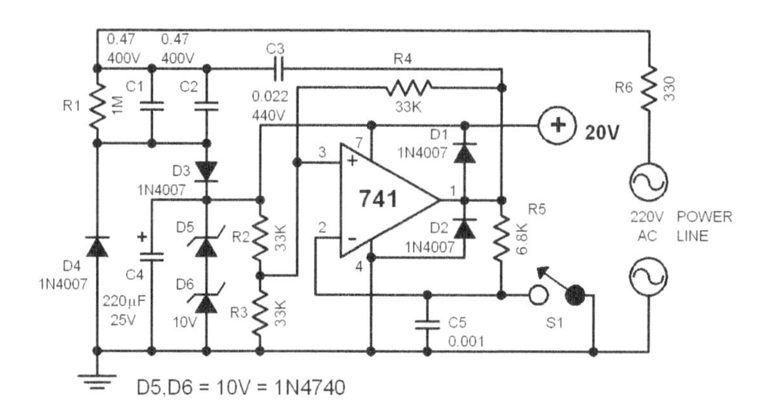

Diagram 86.0 Power Line Remote Control (Transmitter Module)

This circuit is a remote control switch that can turn on or off an appliance via the common household power line. For example, one can turn on a coffe maker from one's bed. The transmitter circuit shown in diagram 86.0 sends a signal of around 36 kHz in the household power line. The receiver circuit shown in diagram 86.1, for its part, monitors the household power line for the presence of this 36 kHz signal. When the receiver detects this 36 kHz signal, it activates the relay. When the 36 kHz signal disappears, the receiver releases the relay.

The transmitter circuit works as simple as its principle. The 741 IC functions as a squarewave generator producing the 36 kHz signal. It is powered through R6, C1 and C2. The four diodes rectify the AC voltage and limit it to 20 volts. Capacitor C4 is the ripple filter. The 36 kHz signal is fed to the household power line via the capacitor C3. Diodes D1 and D2 protect the circuit from being destroyed by the power line. Resistor R1 also offers protection by discharging the capacitors C1 and C2 when the circuit is unplugged from the power line. Both C1 and C2 are rated 0.47µF/400V.

The receiver is made of another 741 opamp IC. It is powered through a small stepdown transformer rated 6V/300mA. The 36 kHz signal is sampled from the household line by the LC network composed of C1,C2,C3 and L1.

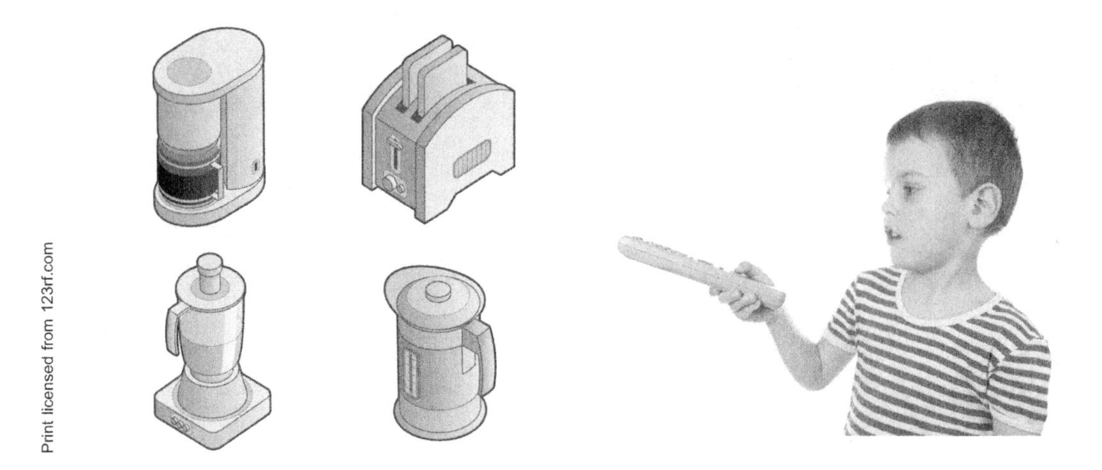

Remote controlling household appliances via the main power lines

The receiver's 741 opamp is dimensioned to work as a 36 kHz bandpass filter. When the 36 kHz signal passes through R1, it it amplified by the opamp and rectified by diode D3 (germanium due to its low threshold). The rectified signal triggers the transistor T1 which in turn triggers T2. The relay closes and any appliance connected to the relay is turned on.

The sensitivity of the receiver circuit can be adjusted through potentiometer P1. The resistor R8 might need to be adapted to the actual current consumption of the appliance. The relay must be rated according to the current consumption of the appliance. For a coffee machine, this can be around 5A.

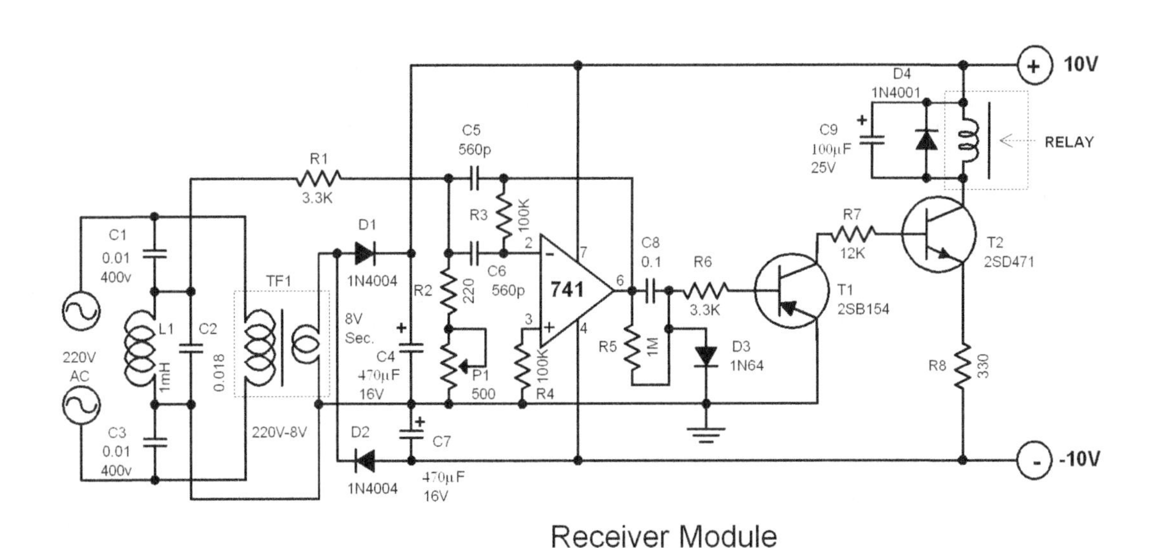

Diagram 86.1 Power Line Remote Control (Receiver Module)

87 WATER LEVEL MONITOR

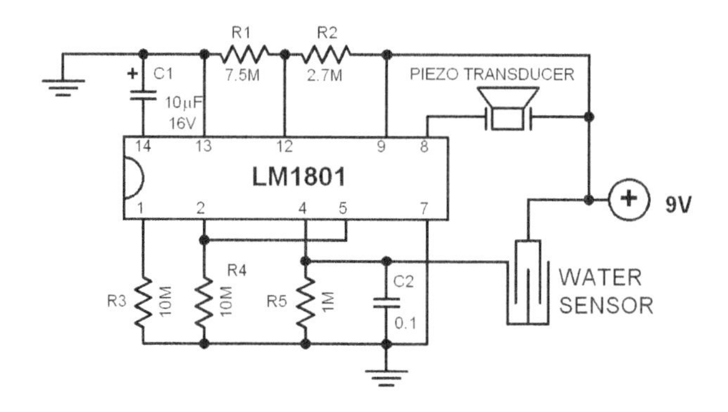

Diagram 87.0 Water Level Monitor

This simple circuit monitors the presence of water in a certain location or container. It is also used as water leak detectors in places like laundry or basements. Conventional self-closing valves or pumps usually work only by high volume of water leaks. They do not function reliably by drops of water leak.

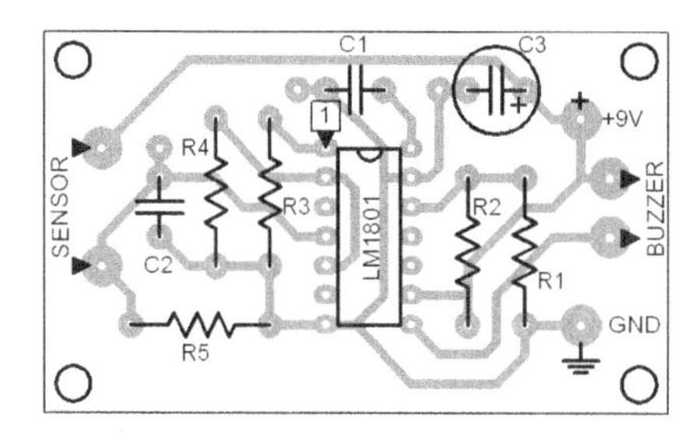

Figure 87.0
Parts Placement Layout

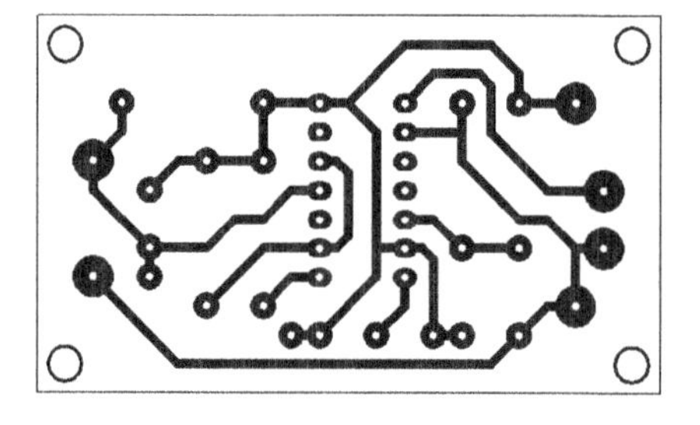

Figure 87.1
Printed Circuit Layout

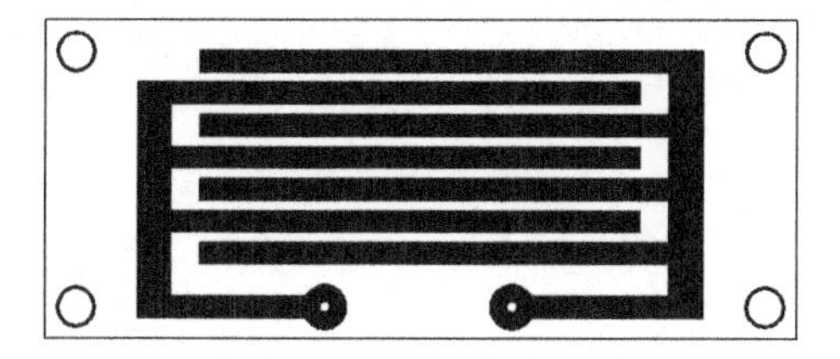

Figure 87.2
The water sensor made of etched PCB

The featured electronic circuit sends out an acoustic alarm when it senses a drop of water leak. It is very simple to build. It is made of a single IC and some passive components. The IC LM1801 is a low-power comparator that can deliver high output current if needed. The reference voltage is set by R4 at pin 2. When water hits the sensor, the reference voltage is overshot and the IC drives the ceramic transducer to beep. It is also possible to connect several sensors to the circuit. The sensor can be easily made out of a small piece of pcb that is etched with the proper pattern. Figure 87.2 is a sample design of a sensor made out of etched pcb board.

The decoupling capacitor C3 (not shown in the schematic diagram) is a 100μF/16V electrolytic capacitor.

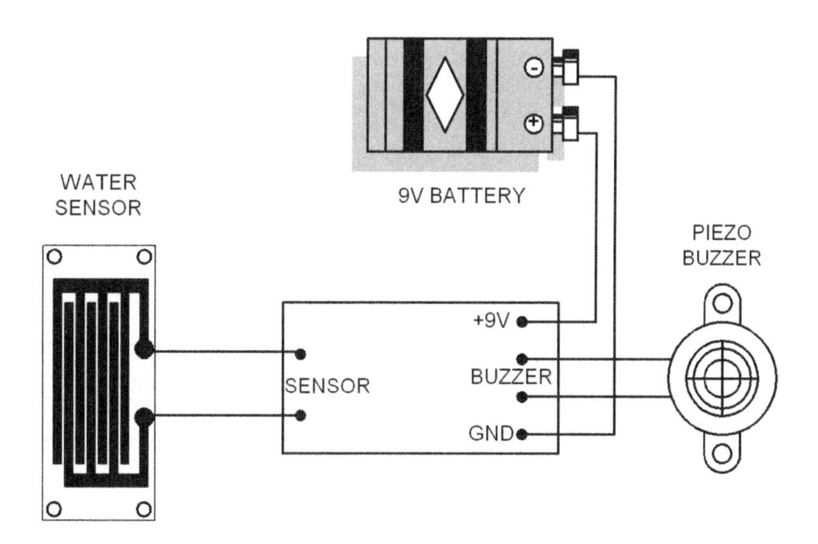

Figure 87.3 External Wiring Layout

88 SIREN

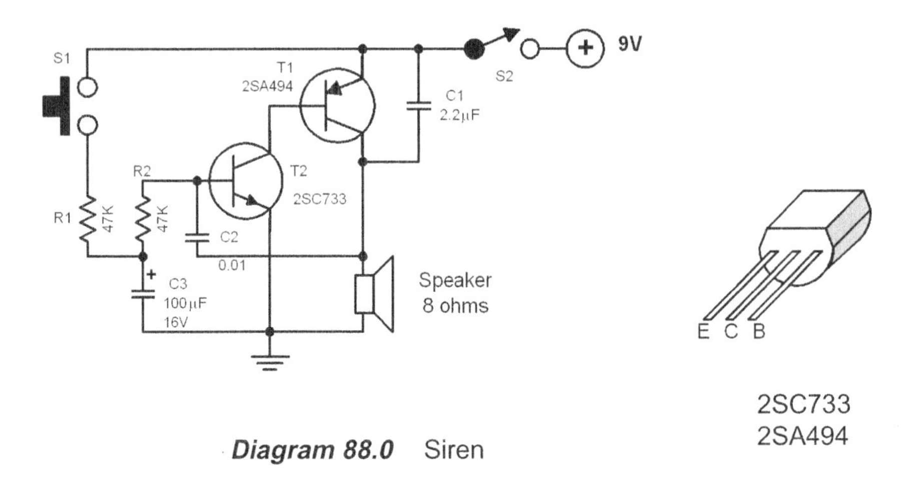

Diagram 88.0 Siren

This circuit generates a tone that sounds very similar to a siren. The generator part of the circuit is made of the combination of PNP and NPN transistors. Together, the two transistors build up a free running multivibrator. If the C2 capacitor was connected to the positive line of the power supply, it would have worked as a constant frequency oscillator.

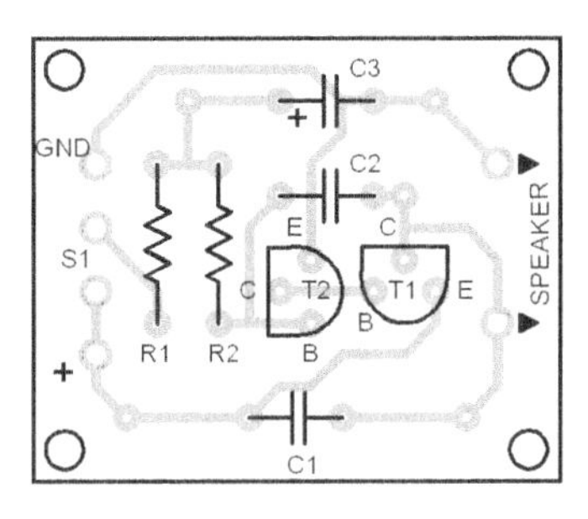

Figure 88.0
Parts Placement Layout

However, we don't want a constant frequency oscillator. We want a siren. So to generate an up and down going signal tone, the resistor R2 is fed from an RC circuit (made of R1 and C1). When the switch S1 is pressed, the capacitor C1 charges via R1 slowly until it reaches the maximum voltage level of 4 volts. This increasing voltage results to a decreasing time constant at the R2/C2 junction. This furthermore results to an increasing frequency of the multivibrator. After the switch S1 is released, the capacitor C1 discharges slowly resulting to a decreasing frequency cycle. Through the combination of the two time constants a sawtooth waveform is generated.

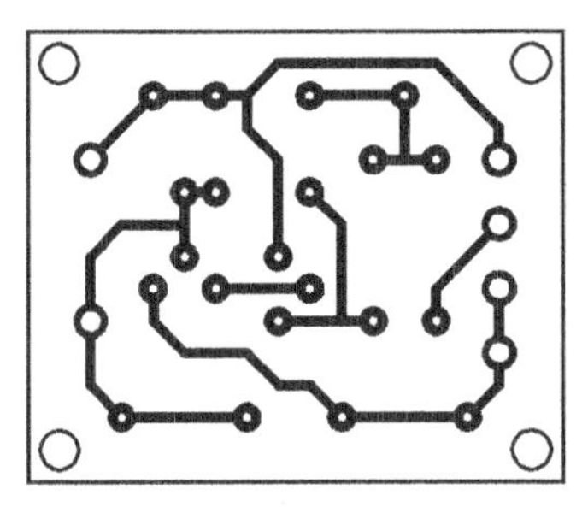

Figure 88.1
Printed Circuit Layout

The signal heared from the speaker will be an increasing or decreasing tone depending on whether the switch S1 is pressed or released.

89 LED OPTOCOUPLER

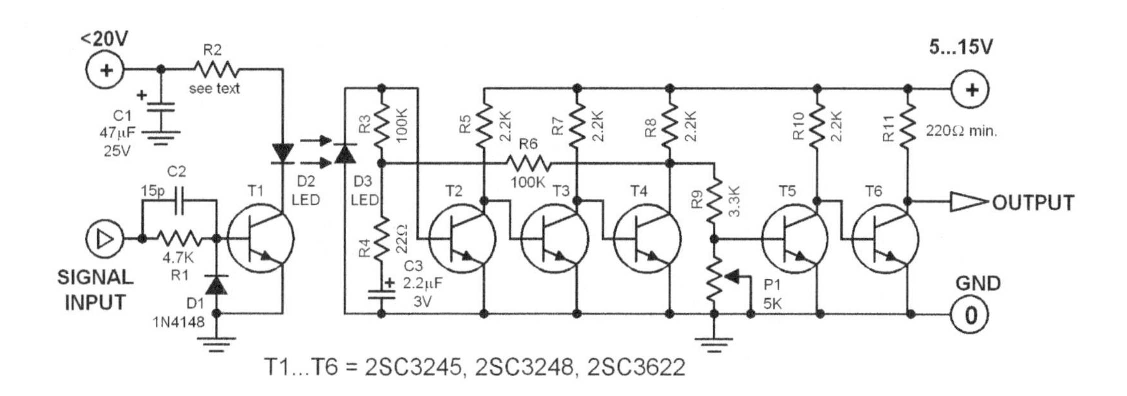

Diagram 89.0 LED Optocoupler

This optocoupler circuit uses two ordinary LEDs as the optocoupler element. When a voltage of 2V is fed to the input, LED2 will generate a voltage that is (after amplified by the following transistors) strong enough to switch on transistor T5. The upper frequency limit is around 38 kHz. Very low pulse frequencies on the other hand are almost meaningless for the circuit.

Eventhough the LEDs are quite insensitive to stray light, it is still a good idea to enclose them with an opaque box. Resistor R2 must be selected basing on the power supply voltage. It can be between 680 ohms and 1K. Resistor R11 must not be lower than 220 ohms. Transistor T6 has an open-collector character.

Parts List

R1 = 4.7K
R2 = 680Ω - 1K(see text)
R3,R6 = 100K
R4 = 22Ω
R5,R7,R8,R10 = 2.2K
R9 = 3.3K
R11 = 220Ω minimum
P1 = 5K trimmer
C1 = 47μF/25V
C2 = 15p/50V
C3 = 2.2μF/3V
D1,D2 = LED
T1...T6 = 2SC3622
(2SC3245)(2SC3248)

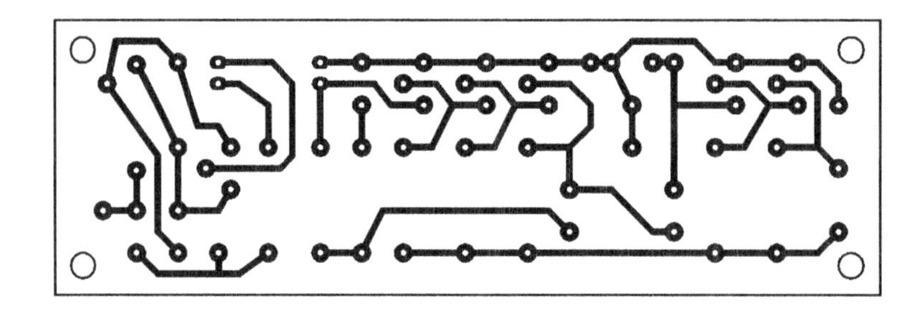

Figure 89.0 Printed Circuit Layout

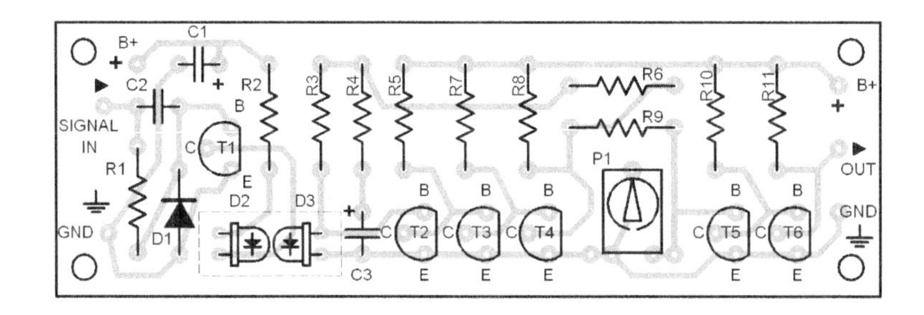

Figure 89.1 Parts Placement Layout

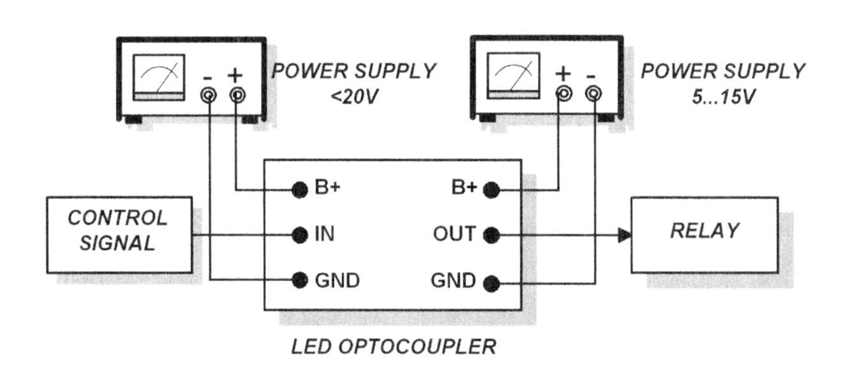

Figure 89.2 External Wirings

90 MUSIC SOUND GENERATOR

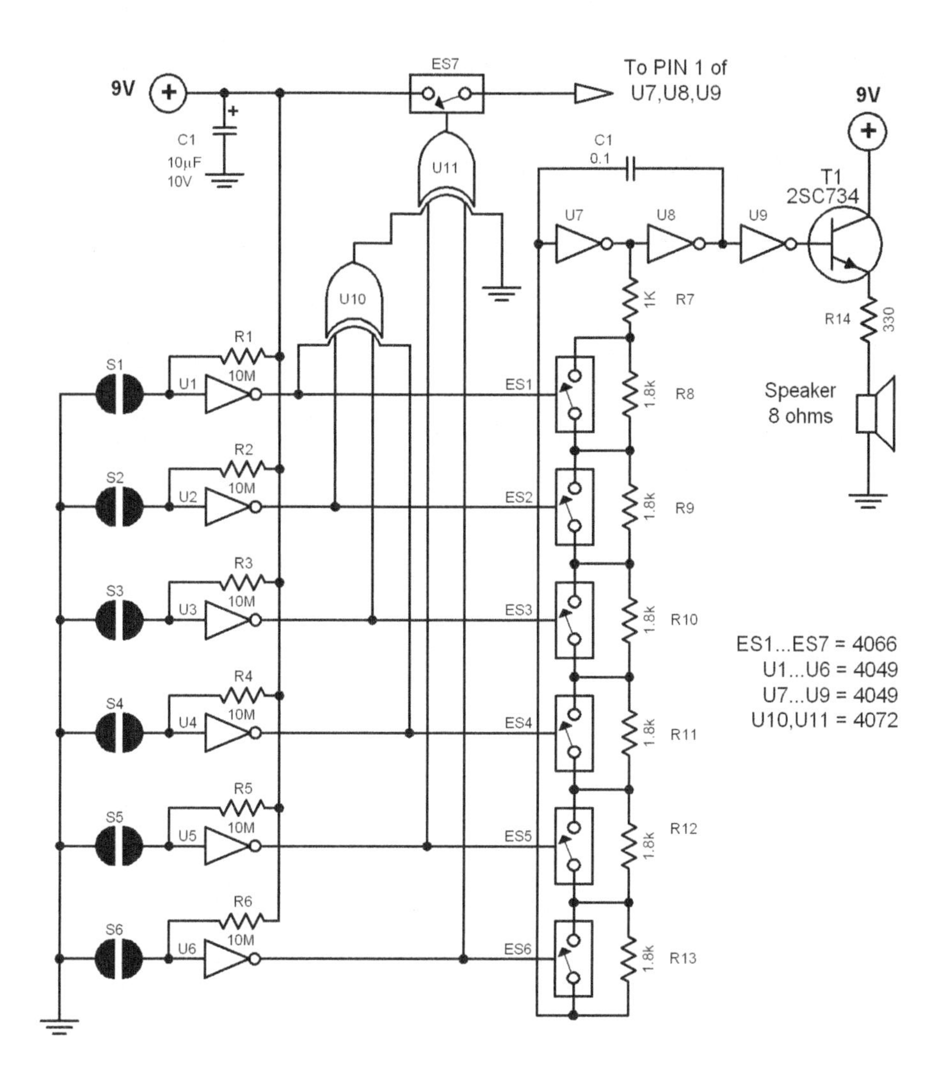

Diagram 90.0 Music Sound Generator

This electronic circuit generates a tone when the buttons (S1 ... S6) are touched. The tone frequency is dependent on the button that is being touched.

91 LOGIC PROBE

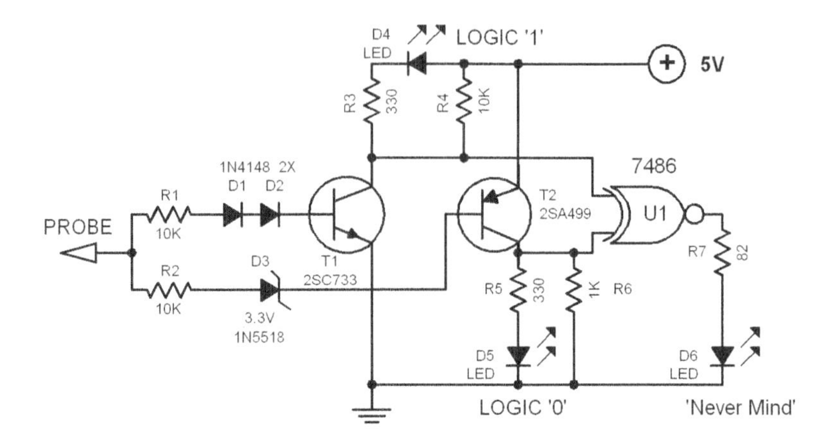

Diagram 91.0 Logic Probe

Logic testers are simple but very helpful devices in testing digital circuits. A logic probe can be designed in many different ways. In this particular design, a combination of discrete and TTL logic components is applied to test different logic levels. This logic tester can test and display three different logic levels: the „0“ and the „1“ levels including the undefined logic state also known as „never mind“.

If the tested voltage level is below 1 volt, the logic tester will recognize it as a logical „0“. In that case, the emitter-collector junction of T2 conducts and the D5 LED lights up optically displaying a „logic 0“. If the tested voltage is between 1 volt and 2 volts, both T1 and T2 does not conduct and the XOR gate U1 receives two dissimilar logic states at its inputs. The XOR output becomes positive and the D6 LED lights up signalling a „never mind“ logical state. The same thing happens by an open input or when testing a blind IC pin.

When the tested point is above 2 volts, the D4 LED lights up signalling a logical „1“ state. Another plus of the featured circuit is the fact that it uses a 7486 IC for the XOR gate. Since this IC has 4 gates inside, the circuit can be expanded to build a four channel logic analyzer.

92 ELECTRONIC POOL

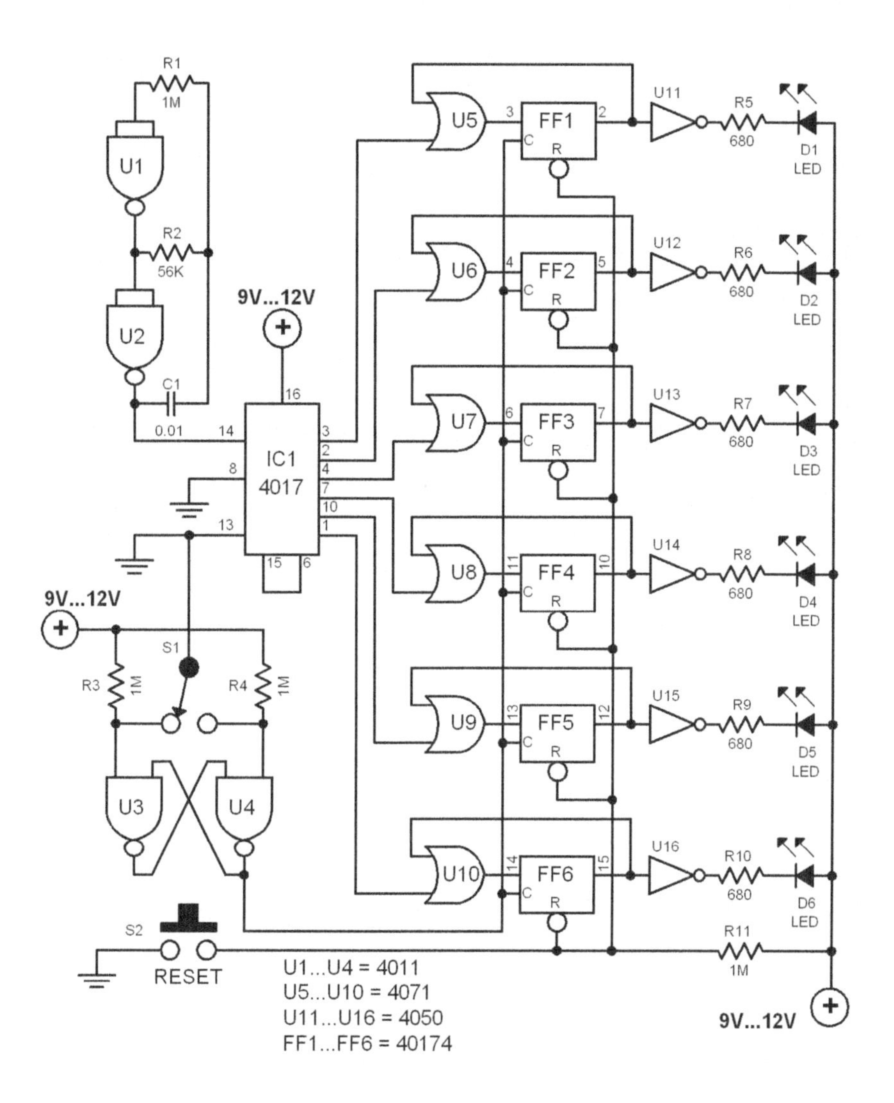

Diagram 92.0 Electronic Pool

The game of pool billard is fast becoming popular. It is even available in electronic version such as the circuit featured here. This electronic pool runs very similar to the real pool billard. The billard balls however are replaced with LEDs in the circuit.

Technically, the electronic circuit is a simple random number generator. After the „Reset" button is pressed, all LEDs will light up. Pressing the „Hit" button will start the random generator that results to either one of the LED is turned off or all the LEDs stay lighted. The LED that is turned off symbolizes a ball that fell into the side pocket.

There are two possible rule variations for this game. In the first case, the player choose to sink all the LEDs (balls). The player who manages to do this with the least number of „hits" (pressing S1) wins. In the second case, the game is limited to two players. One player begins. When he hits a red LED, he must hit all other red LEDs to win. The green LEDs automatically become the other player's balls. As long as the first player hits his color, he keeps playing. Only when he does not turn off his LED or turns off the opponents LED will he stop playing and the opponent comes to play his turn. Once three balls of the same color have been hit, the player of this color wins the game.

The electronic circuit works this way: The switch S2 sets all flip flops FF1 to FF6 to „zero" logic and all LEDs light up. The multivibrator made of U1 and U2 feeds a constant clock frequency of 800 Hz to the 4017 counter. The counter delivers successive „1" levels to the output.
The OR gates U5 to U10 connect the counter outputs to the N-inputs of the flip flops. A short press to S1 feeds a clock pulse to all flip flops. Every counter output that has a „1" logic at this moment will set the flip flop via the OR gate when it is not set yet. The LED that is connected to it will be turned off. The feedback loop of the Q output to the D-input via the OR gate enables the flip flop to stay set even if additional clock pulses arrive.

The 4050 IC can be replaced with a 4049 that contains six inverting drivers. This will reverse the function of the LEDs. After a reset of S2, all LEDs will turn off. A lighted LED will symbolize a ball that sunk into the side pocket.

The power supply for the circuit must deliver a maximum current of 100 mA.

Parts List:

Resistors:
R1 = 1M
R2 = 56K
R3 = 1M
R4 = 1M
R5,R6,R7,R8 = 1M
R9,R10,R11 = 1M

All resistors are ¼ watts unless otherwise specified.

Capacitors:
C1 = 0.01 µF

IC's:

IC1 = 4017
U1,U2,U3,U4 = 4011
U5,U6,U7,U8,U9,U10 = 4017
U11,U12,U13 = 4050
U14,U15,U16 = 4050
FF1... FF6 = 40174

93 PHOTO FLASH TRIGGERS

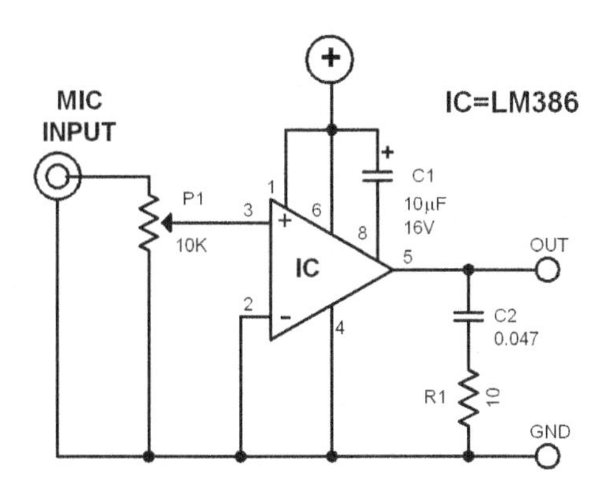

Diagram 93.0 Microphone Amplifier

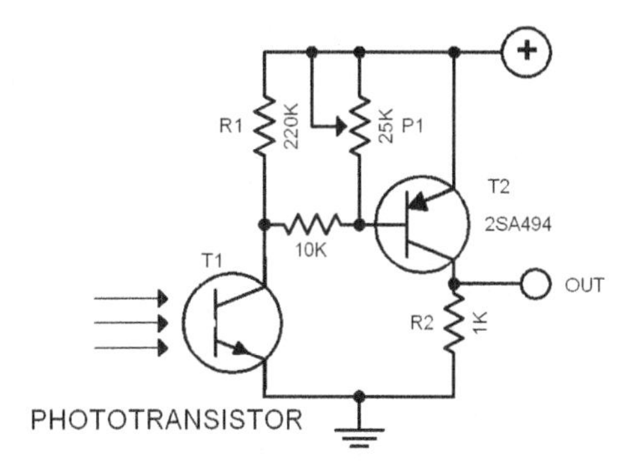

Diagram 93.2 Light Trigger

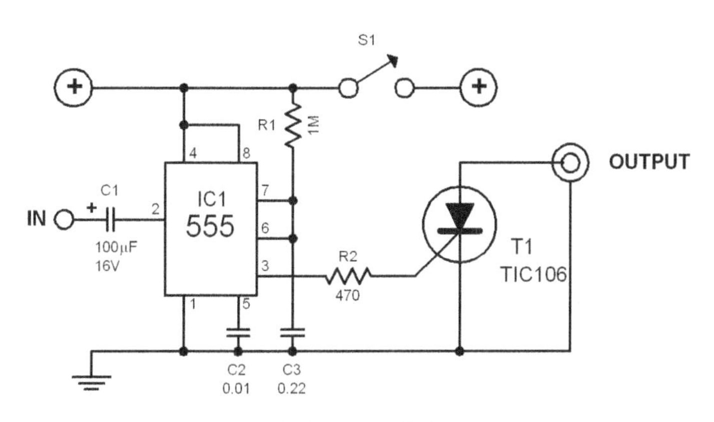

Diagram 93.1 Microsecond delay

The featured circuits were designed to help a hobby photographer in shooting photos of fast events that produce (or accompanied by) a loud noise or a bright flash of light. Typically, commercial versions of these circuits are very expensive and therefore mostly reserved for commercial photo enthusiasts.

The first circuit shown in Diagram 93.0 is applicable for an event that produce (or is accompanied) by a loud noise. Potentiometer P1 is set to what level of noise must the flash strobe be triggered. The output of this circuit is connected to the „trigger" or strobe input of the flash. A crystal or piezo microphone is connected to the input of the circuit.

The power supply is not critical. A common 9 volts battery can be used. The circuit consumes around 30 mA.

The IC1 of the first circuit is a common LM386 audio amplifier. It is wired here as a microphone amplifier with a maximum gain of 200. IC2 of the microsecond delay on the third circuit, shown in Diagram 93.1, is 555 timer IC that is wired as monoflop. If a negative pulse is present at its input (as a result of a loud noise), the monoflop triggers. The monoflop output at pin 3 triggers the thyristor T1. This thyristor further triggers the thyristor inside the flash.

The first circuit must be connected to the microsecond delay circuit by joining the power supply lines together, connecting the output line of the microphone circuit to the input line of the 555 circuit and joining the ground lines together. See Figure 93.0.

If the triggering event is not a noise but a flash of light, the microphone amplifier can be replaced with the phototransistor circuit as shown in the third diagram 93.2. In this case, when a ray of light hits the phototransistor T1, the circuit triggers the flash.

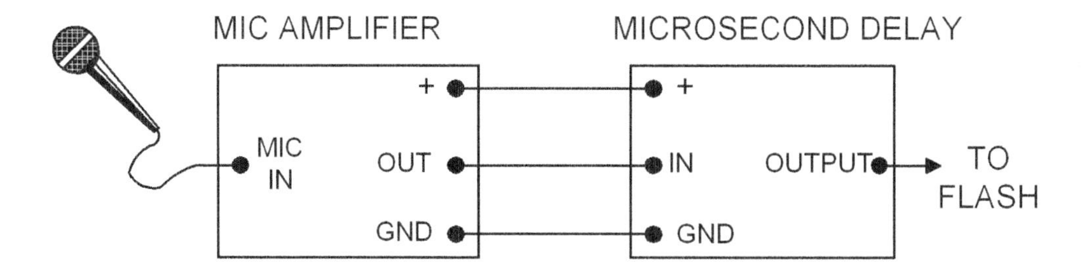

Figure 93.0 Joining the two circuits

94 SIGNAL METER

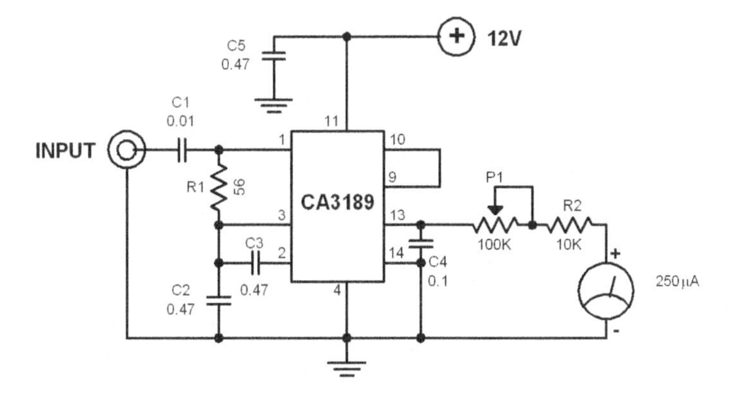

Diagram 94.0 Signal Meter

The CA3189 IC integrates a symmetrical limiter, a phase demodulator, an audio amplifier, and a logarithmic detector-amplifier. The logarithmic detector-amplifier is the most interesting part of the IC for the above featured circuit. It can be used to enhance the „logarithmic signal meter" of shortwave receivers to make the signal reading more accurate.

Building the circuit is very simple. It must be built in such a way that it will work up to 30 MHz. To do that, follow these guidelines: connect the input to a 50 ohms line; make the circuit wiring as short as possible; and the input signal must be shielded.

95 CRYSTAL TESTER

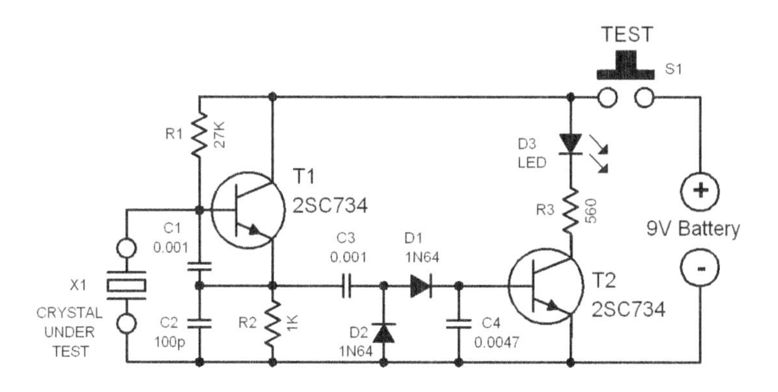

Diagram 95.0 Crystal Tester

The idea for this circuit sprung out of the need of testing a large number of oscillator crystals lying unused in a big hobby box. Testing the crystals one by one without the appropriate device would have been very slow and a time consuming task. Commercial crystal testers are however very expensive, that is why this simple electronic circuit was born.

The transistor T1 and the crystal to be tested together makes a complete crystal oscillator. The capacitors C1 and C2 work as a capacitive voltage divider that is connected to the transistor T1. If the crystal being tested is intact, the circuit oscillates. The sinus wave oscillation voltage is fed to the rectifier circuit (D1,D2) and filtered by capacitor C4. With an intact crystal, the DC voltage at the base of the transistor T2 is high enough to cause the transistor to conduct. The LED lights up signalling that the crystal is good.

The electronic circuit can be used to test crystals with frequencies from 100 kHz up to 30 MHz. The current consumption is low: around 25 mA.

96 PULSE DUTY CYCLE METER

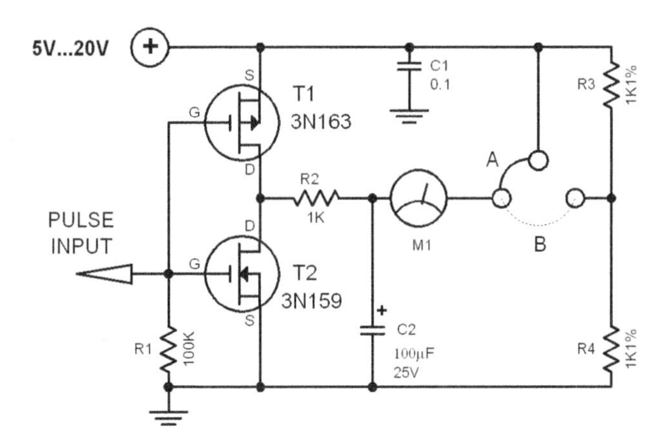

Diagram 96.0 Pulse Duty Cycle Meter

This is a simple test circuit to measure pulse duty cycle ratios. In normal cases, the duty cycle of square wave signals are tested using a pulse counter or oscilloscope. However, it can be done easier by using this circuit and a simple voltmeter.

The two VMOS-FET transistors in the circuit are toggled by the incoming pulses. The AC voltage is averaged by the R2/C2 network. The resulting DC voltage can be measured using a voltmeter in two different ways. First, it can be tested for the duty cycle in % by using the A jumper. Second, it can be tested if the pulse ratio is symmetrical (meaning 50%) by using the B jumper. In the second case, it is necessary to use a voltmeter with a center scale zero point. A digital voltmeter can be used in place of the analog voltmeter.

When the duty cycle is 50%, the DC voltage will be one half of the supply voltage. Since one half of the supply voltage is at the other side of the circuit through R3/R5, no current flows through the voltmeter and its pointer stays in the mid scale (zero point).

If the A jumper is used, it is necessary to use a regular voltmeter with a scale of 0...10V (if the power suppy is 10 volts). The duty cycle can be directly read by just adding a „0" at the end of the measured voltage. For example, if the meter reads 3, then the duty cycle is 30%.

Very important:

> The maximum „0"-level must be 0.8 volt and the minimum „1"-level must be 5 volts. Otherwise, both FETs will conduct and short the supply line to the ground line! The input level must not exceed the power supply level.

In choosing the volt meter display, the maximum deflection level must equal the power supply voltage. The internal resistance of the meter must have a minimum of 100 kiloohms.

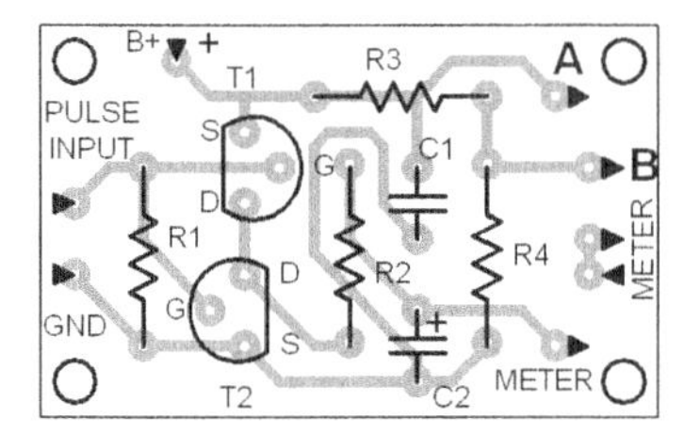

Figure 96.0
Parts Placement Layout

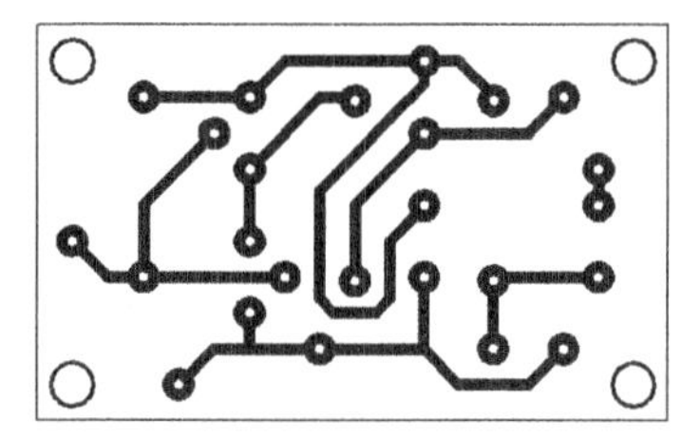

Figure 96.1
Printed Circuit Layout

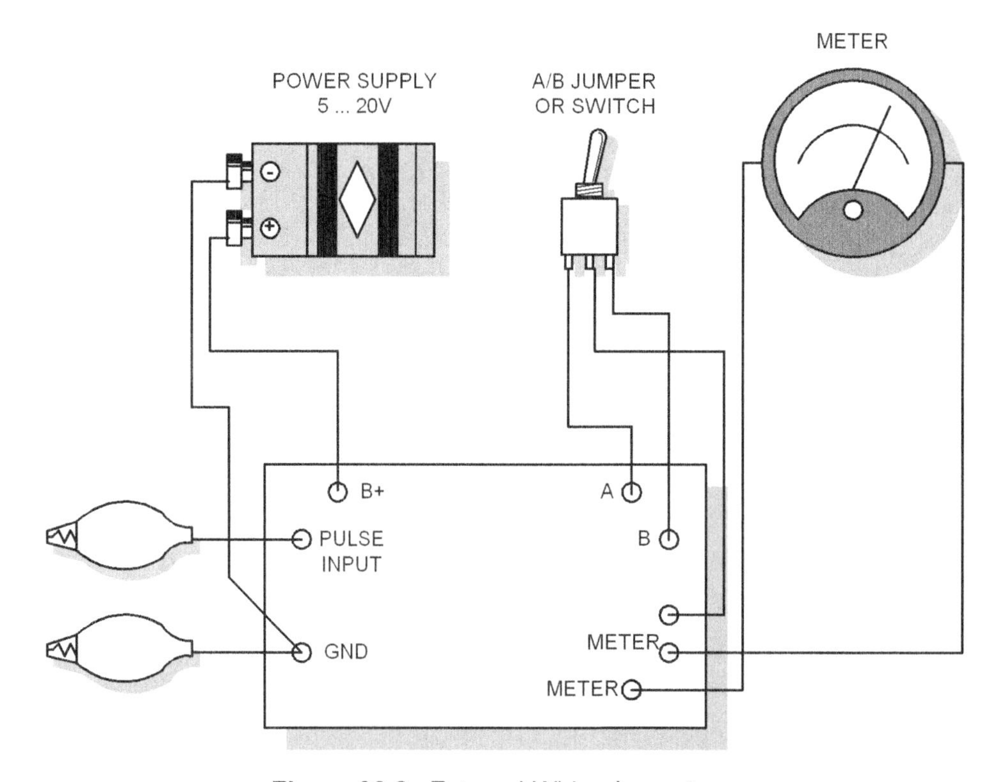

Figure 96.2 External Wiring Layout

97 LIQUID LEVEL SENSOR

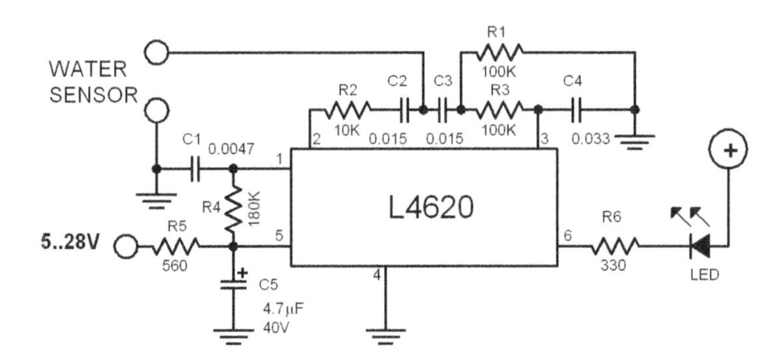

Diagram 97.0 Liquid Level Sensor

A circuit that sounds off an alarm when a liquid has reached a certain level can be easily constructed using a single IC L4620 from SGS and few external components. The featured circuit oscillates at 1.6 kHz. This signal is divided into 32. The resulting 50 Hz clock signal is fed to the sensor through pin 2. The sensor is nothing more than a simple moisture dependent resistance. It is usually made of two wires in the liquid which either shorts the pin 3 to the ground or not. The RC components between pin 3 and pin 2 work as a bandpass filter with a center frequency of 50 Hz. This bandpass filter is, however, needed only in critical cases. The capacitors C2 and C3 can be jumped in most applications.

The sensor interface receives a 200 Hz signal and compares the voltage at pin 3 with a reference voltage which is dependent on pin 2. The reference voltage is within 0.2 and voltage of pin 2 when pin 2 is low. It is within 0.4 volts and voltage of pin 2 when pin 2 is high. Once these thresholds are exceeded, the sensor interface passes the information to the rest of the circuit.

To sound an alarm when the threshold is exceeded, pin 8 must be high. On the other hand when pin 8 is low, the alarm output is active. The alarm will not trigger immediately. It triggers only when the alarm situation (liquid level is reached) remains constant at the sensors for 10 seconds (pin 7 is low) or for 20 seconds (pin 7 is high).

This electronic circuit can consume up to 300 mA during an alarm. The current is dependent on the voltage supply. At 5 volts supply, the current is typically around 6 mA.

98 Digitally Controlled Trigger

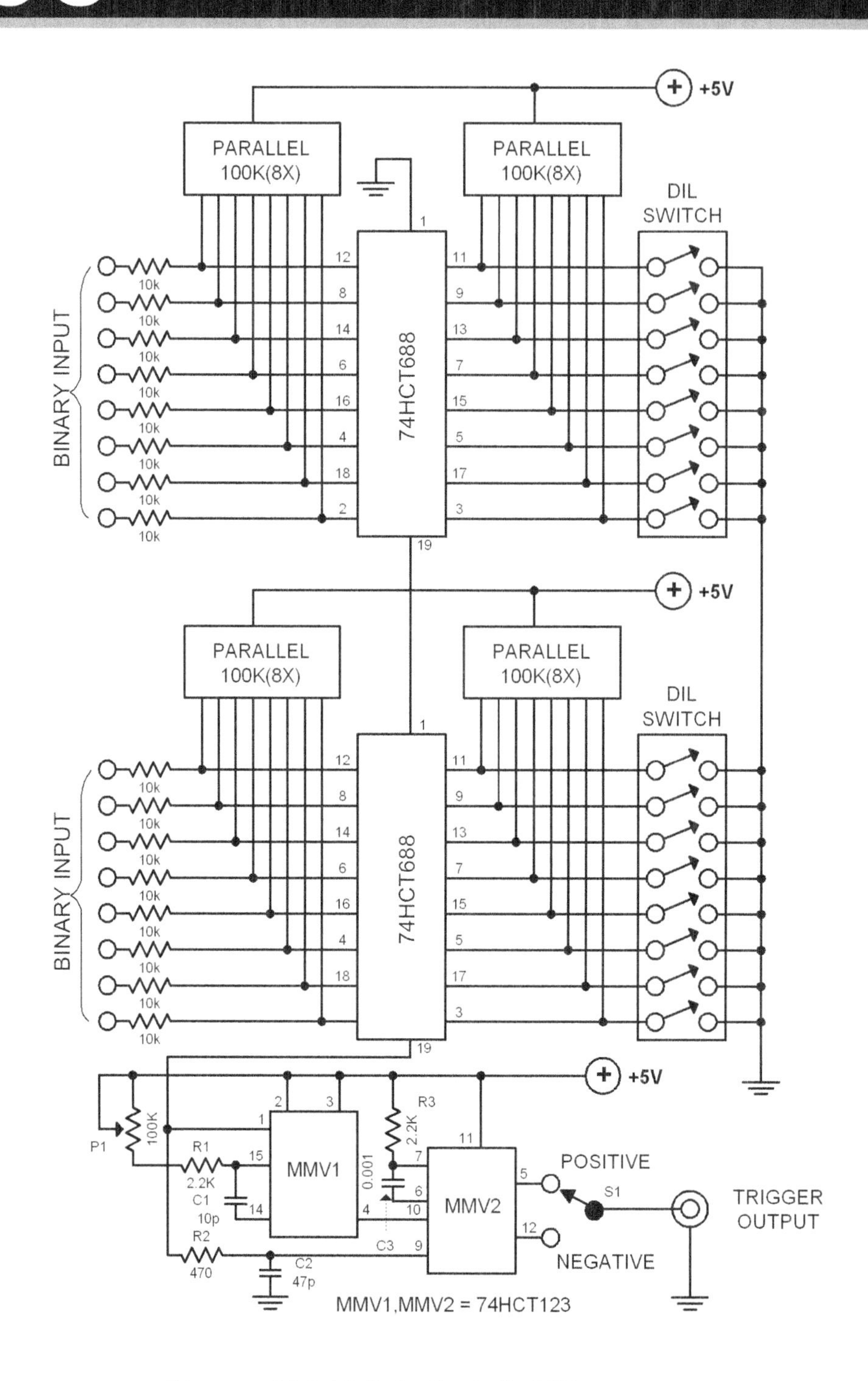

Diagram 98.0 Digitally Controlled Trigger

This circuit was initially designed to trigger an oscilloscope when a certain preprogrammed binary code appears at the circuit's input lines. It can be used for triggering other devices as well.

Open inputs are seen as logical „1" or „high" due to the pull up resistors. The integrated ICs 74HCT688 compare the bit code at the 16 input lines with the bit code that was programmed through the DIL switches. When these two bit codes are the same with at least 100 nS, the pin 19 of IC1 (the lower 74HCT688) will become a logical „0". This will in turn trigger the MMV2 causing the Q output to deliver a 0.1 to 1.5 microseconds long negative pulse. The pulse length depends on the position of P1. However, if the trigger bit code disappears from the input lines during this time, the MMV will not be triggered. It is best to use a logarithmic potentiometer for P1 to enable exact setting of shorter times.

The output of MMV2 triggers the MMV1. The time constant of MMV1 is set to 1 microsecond with R23 and C3. The switch S3 selects either the positive(Q) or negative (-Q) signal to trigger the oscilloscope or any similar device.

The capacitor C4 (not shown in the schematic diagram) is rated 0.1µF/60V ceramic. The long strips labeled 100K(8x) are parallel resistor strips containing eight 100K resistors each. The resistors group marked 10K's are individual resistors rated 10K each.

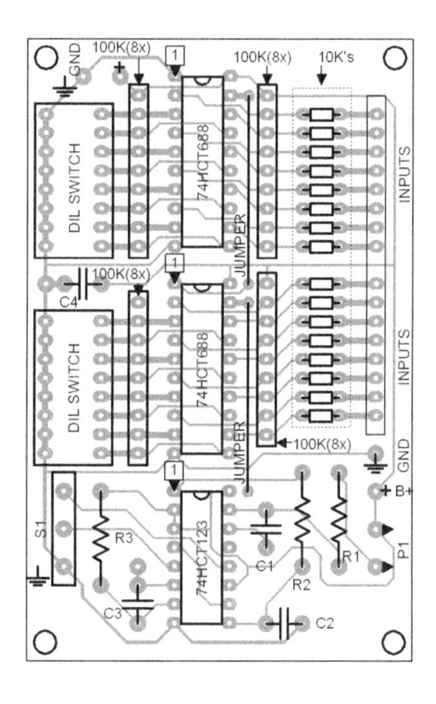

Figure 98.0
Parts Placement Layout

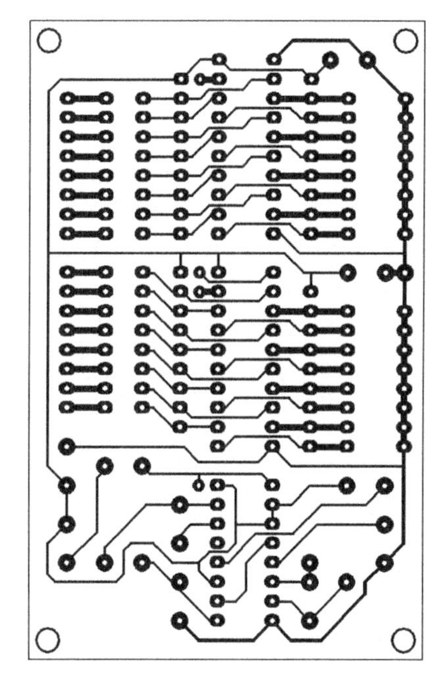

Figure 98.1
Printed Circuit Layout

99 TEMPERATURE MONITOR

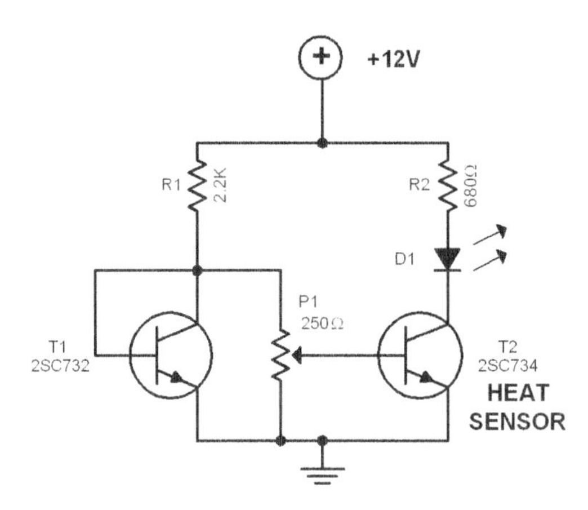

Diagram 99.0 Temperature Monitor

High power amplifiers usually dissipate large amount of heat. It is therefore practical to monitor the level of heat from its power transistors and heatsinks and if needed, to automatically turn off the amplifier to avoid damage to the vital components. This can be done by using simple temperature monitors such as the circuit featured here. Since it is not required to monitor the temperature by strict degree resolution, this simple monitor will work very well.

The electronic circuit works by comparing the voltage drop of a „cold" diode (T1) with the base-emitter voltage of a „warm" transistor (T2). The transistor must be attached physically closest to the heat source. Ideally it should be attached to the heatsink of the power transistors. The diode (T1) must be positioned away from the heat source to ensure that it is always at room temperature. The circuit „measures" the heat difference between the transistor and the diode. The diode (T2) is connected to the power supply via R1. The base of T2 is connected to the anode of the diode via the potentiometer P1. The LED D1 and resistor R2 is connected in series to the collector of T2. Transistor T2 should not conduct as long as the temperature level being monitored is below the set threshold. The base-emitter voltage of T2 decrease by 2 mV per °C. Once this voltage is below the level set by P1, the transistor T2 conducts and the LED begins to light up. By a slow increase of temperature one will notice a dim light from the LED. This way, the user gets an „analog" indication of the temperature.

The values of R1 and R2 are dependent on the power supply voltage. The values can be computed using the following formulas:

R1 = (Ub/V - 0.6) / 5 kiloohm

R2 = (Ub/V - 1.5) / 15 kiloohm

For example: by 12 volts power supply, the R1 is 2.2 kiloohms while R2 is 680 ohms. The maximum current consumption of the circuit when the LED lights up is 20 mA. Take note that the transistor T2 must not become hotter than 125°C.

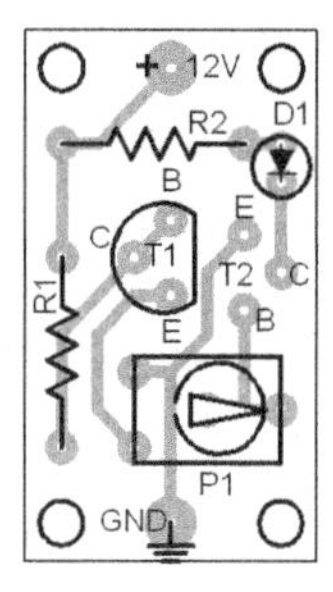

Figure 99.0
Parts Placement Layout

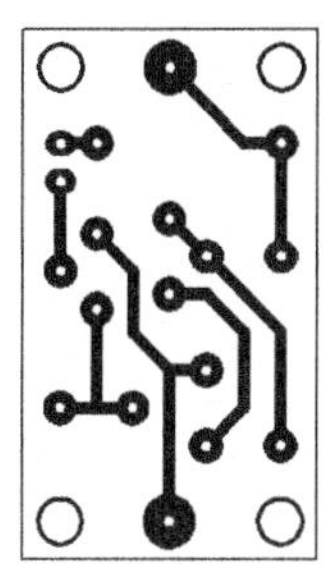

Figure 99.1
Printed Circuit Layout

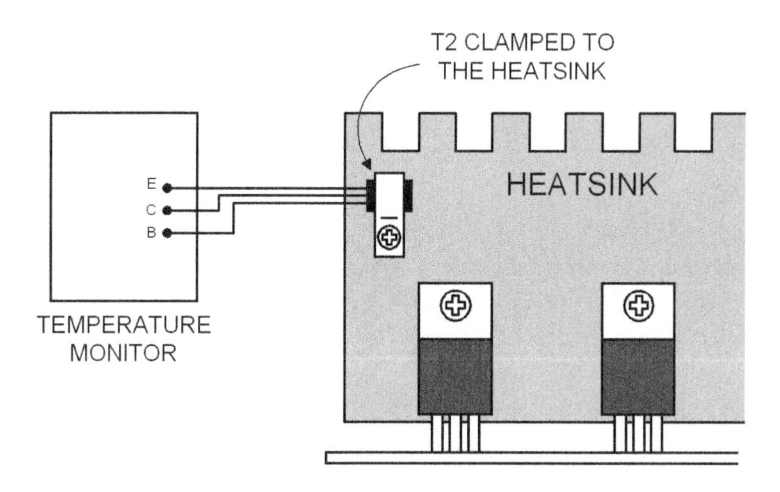

Figure 99.2 External wiring of the T2 temperature sensor and placement to the heatsink being monitored

100 MICROAMPERE METER

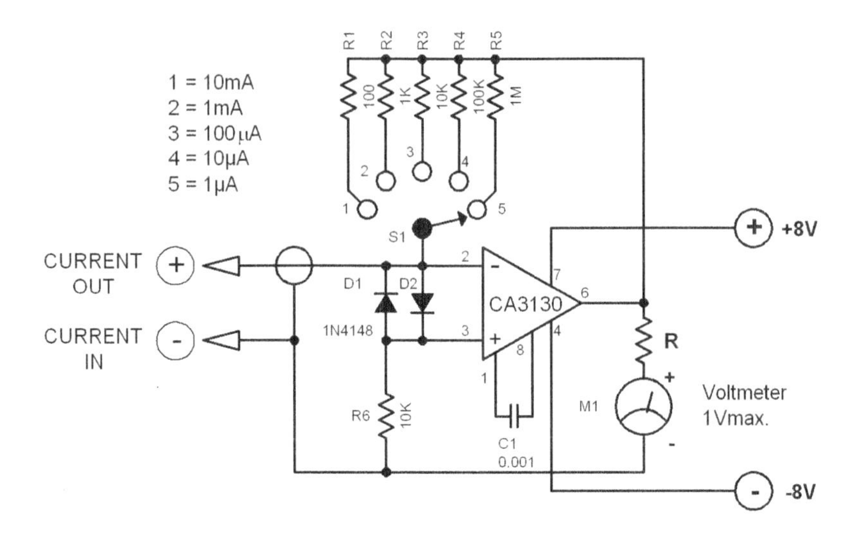

Diagram 100.0 Microampere Meter

This simple device can help in measuring small currents (typically in microampere) in five ranges: from 1 microampere up to 10 milliamperes. The theoritical input resistance of the tester circuit is 0 ohms. Considering that a zero potential is always present at the non-inverting input of the opamp, the circuit works this way: the current being measured Ix shifts the input voltage resulting to an output voltage with an inverted polarity. The output voltage of the opamp CA3130 is proportional to the measured current Ix. By selecting the proper feeback resistors through S1, the output voltage by full meter deflection is 1 volt in all measuring ranges.

The value of the series resistor R must be selected for the particular meter being used. For example: if a 1 mA meter is used, the total resistance (the sum of the resistor R and the coil resistance Ri of the meter) must be 1 kiloohms. If a 100 microampere meter is used, the total resistance must be 100 kiloohms. If needed, a potentiometer can be used for the R.

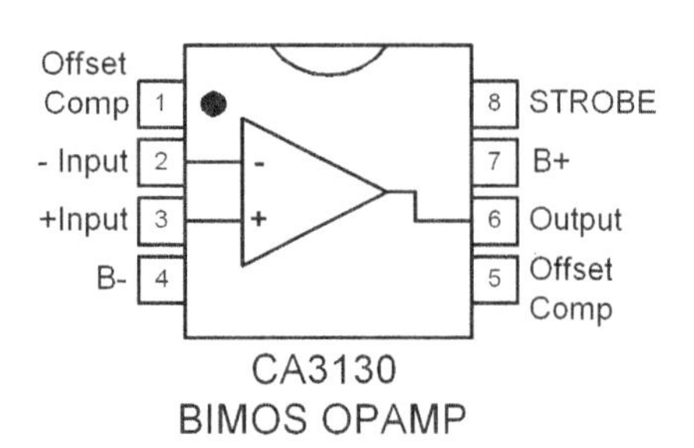

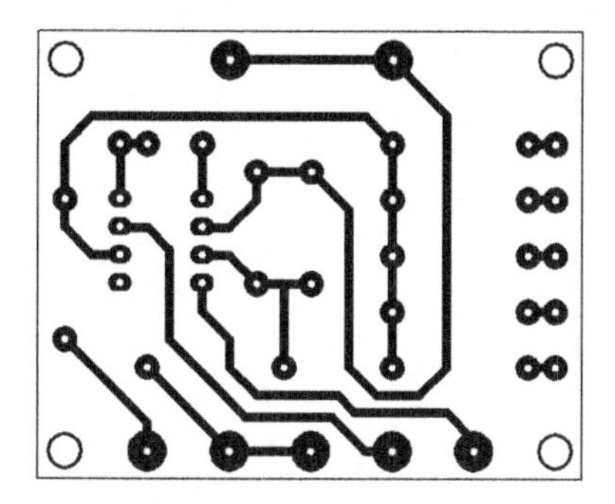

Figure 100.0
Printed Circuit Layout

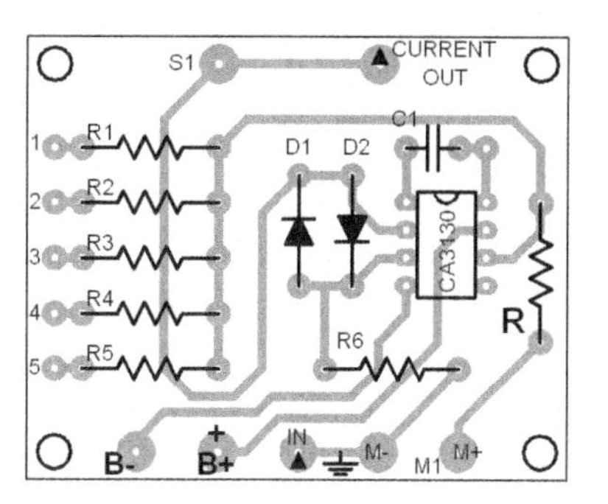

Figure 100.1
Parts Placement Layout

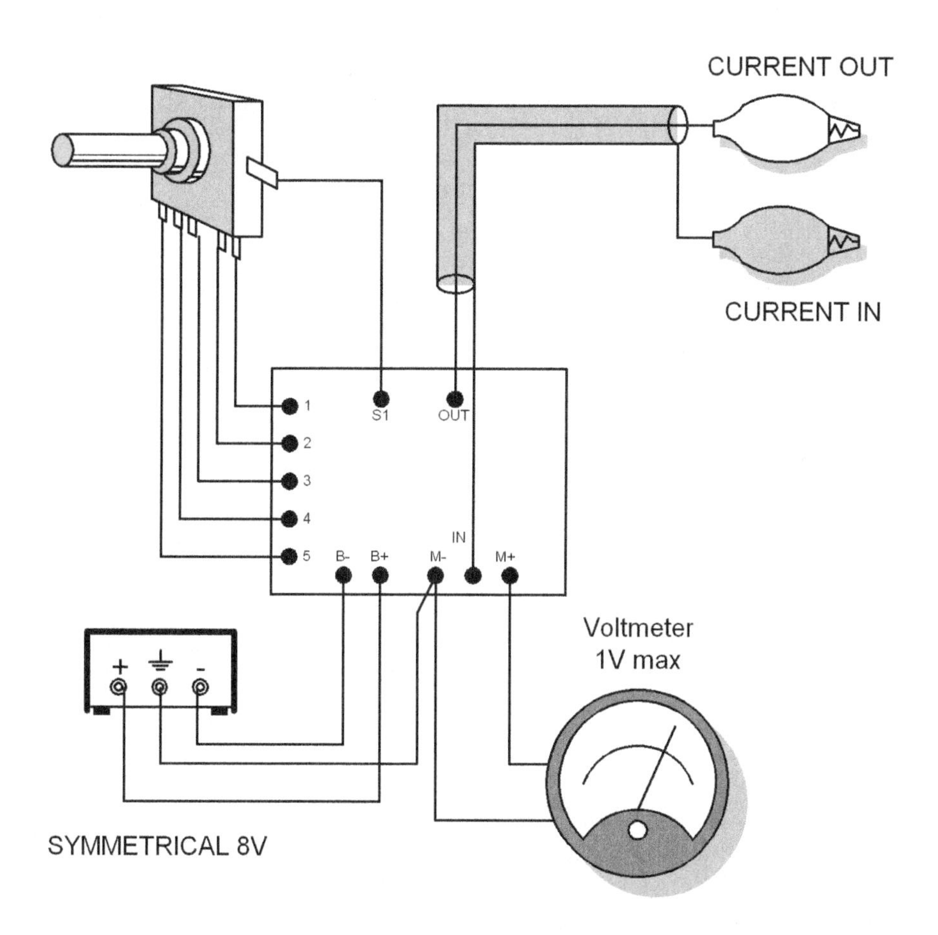

Figure 100.2 External Wiring Layout

101 TRANSISTOR TESTER

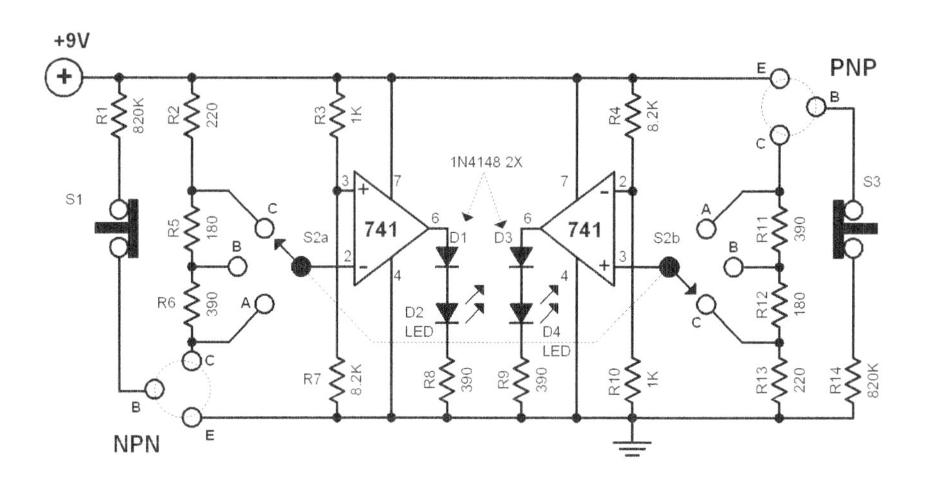

Diagram 101.0 Transistor Tester

This tester circuit helps in speeding up the testing of boxloads of transistors. Not only that it tests whether the transistor is good or defective, it also shows whether that transistor belongs to group A (current gain of 140 - 270), group B (current gain of 270 - 500), or group C (current gain of 500 and higher).

An NPN transistor gets tested in the following way: The transistor is plugged into the „NPN" labeled transistor socket and S2 is switched to „C" position. If the LED D2 lights up, then the transistor belongs to group C. If D2 stays unlit, then S2 gets switched over to position B. If D2 remains unlit then S2 gets switched to position A. If the LED D2 stays unlit in all three positions, then the transistor is either defective or its current gain is less than 140. With low power transistors, a current gain of less than 140 means it is useless. A second test phase is to press the S1 button thereby breaking the current to the transistor base. In a normal case (D2 is lit), the LED D2 should turn off once S1 is pressed. Otherwise if the D2 remains lit, it means that there is a short circuit between the collector and emitter of the transistor.

How the electronic circuit works: A current of around 10 microamperes flow to the base of the transistor through R1. With a good transistor, this base current causes a collector current to flow through R2, R3 and R6 producing voltage drops that are selected by S2. Depending on the position of S2, a sample of this voltage drop is fed to the 741 opamp wired as comparator and compared with a fixed reference voltage. The right part of the circuit is identical to the left part but it is used to test PNP transistors. A 9-volt battery is sufficient to power the circuit.

102 ACOUSTIC OHMMETER

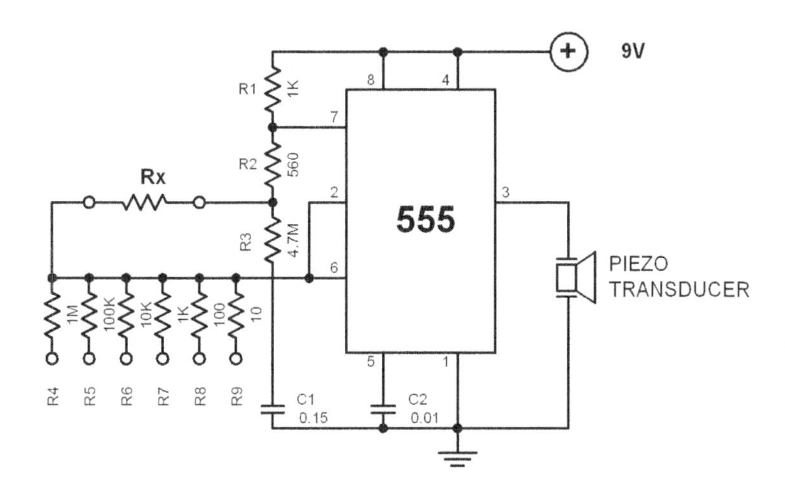

Diagram 102.0 Acoustic Ohmmeter

Determining the value of a resistor by acoustic method can be helpful in special situations.

Sometimes it is useful to have the capability to know in what resistance range does a resistor belong. It is specially useful when one is working under an artifical light that makes it difficult to determine the color code of the resistor.

The featured circuit is an astable multivibrator made of the common 555 timer IC. The output of this oscillator is connected to a piezo transducer. The frequency of the oscillation is directly proportional to the resistance value of the resistor under test Rx (the unknown value).

If the Rx is 0, the output frequency is above 4.5 kHz. If the inputs are open, the output frequency is around 2 Hz. Using the comparison resistors R4 to R9, one can easily determine the resistance group of the resistor being tested. In fact, after some practice and good hearing, one can determine the resistance group without using the comparison resistors.

103 WATER THERMOMETER

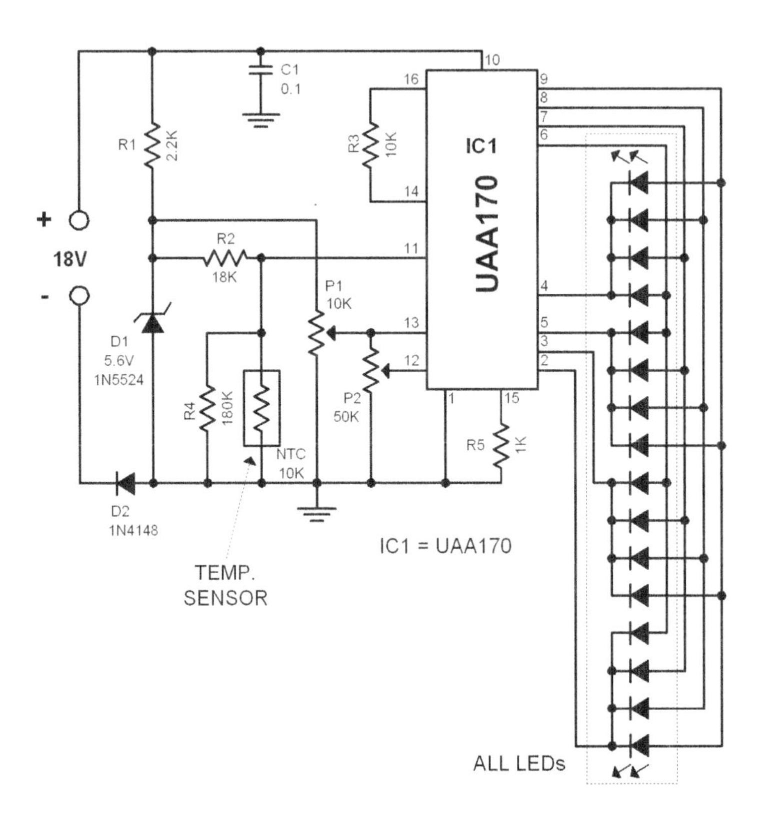

Diagram 103.0 Water Thermometer

This simple to construct device is initially intended to be used in sports applications like for example a fishing contest. A sensor measures the water temperature at a certain depth. A scale made of 16 LEDs shows the water temperature. With the knowledge about water temperatures and what types of fishes prefer what temperature, the fisherman can select the right bait.

The electronic circuit is made of the UAA170 IC with additional external components for the relevant temperature spectrum of 4 to 19 °C (39 to 66 Fahrenheit). P1 is used to set the threshold level for the top LED. P2 is used to set the threshold for the bottom LED. One can also use different colors for the LEDs to visually differentiate the temperature levels.

The input voltage is created through the voltage divider R2/NTC. The NTC is the temperature sensor. Its resistance is reverse proportional to the water temperature. If the water temperature is higher, the NTC resistance is lower resulting to a higher input voltage.

Constructing the sensor: shrink wrap the NTC and the soldered cable with a heat shrinkable tube then insert it into a brass tubing. Finally, pour epoxy into the brass tubing letting the NTC head protruding a little bit out of the tubing.

The entire circuit with the 9-volt battery must be housed in a waterproof box. The cable that connects to the sensor can be marked for the different water depths.

With the knowledge about water temperatures and what types of fishes prefer what temperature, the fisherman can select the right bait.

This page intentionally left blank

APPENDICES

Specifications of the transistors used in the projects

Descriptive Part of the Table:

Type
The original type designation has been taken over directly from the manufacturers, with the abbreviation of the manufacturer added in brackets only in those cases in which different manufacturers used the same type designation.

Mat.
The materials used are abbreviated as follows:

Ge	Germanium
Mos	MOS technology (metal oxide silicon)
Si	Silicon
V-MOS	Vertical MOS technology

Pol.
The polarities used are abbreviated as follows:

npn	NPN structure
n-ch	N channel type (FET)
n-p	More than one transistor with different polarities in one case
pnp	PNP structure
p-ch	P channel type (FET)

Abbreviations used in the following table:

A	Antenna amplifer
AGC	Regulating steps
AF	AF range
AM	AM range
CATV	Broad band cable amplifier
CB	CB-radio
CTV	Colour television application
chop	Chopper
Darl	Darlington transistor
dg	Dual Gate (FET)
double	Paired types
dr	Driver stages
dual	Dual transistor (differential amplifier)
end	Final stages
FET	Field-effect transistor
FET-depl.	Field-effecttransistor, depletion type
FET-enh.	Field-effect transistor, enhancement type
FM	FM range
fs	Fast switch
HD	Horizontal deflection
hi-rel	high reliability
Idss	Drain source short-circuit current (FET)
IF	IF applications
in	Input stages
iso	insulated
ln	Low noise
min	Miniaturised version
mix	Mixer stages
nixie	Digital display tube

osc	Oscillator stages
pow	Power stages
radiation	Aerospace applications (radiation-proof)
RF	RF range
s	Switch
SMP	Switch-mode power supply
SSB	Single side-band operation
Stabi	Stabilisation
sym	Symmetrical types
TV	Television applications
Ugs	Gate source voltage
UHF	UHF range > 250MHz
uni	Universal type
Up	Pinch-off voltage (FET)
VD	Vertical deflection
VHF	VHF range 100-250 MHz
Vid	Video output stages
+Diode, +di	With integrated diode
../..ns	turn-on/turn-off time

Data Part of the Table:

In the case of the ratings, either average values are quoted (< = max.) or lower (> = min.) guaranteed values. As a rule apply at 25°C, unless otherwise indicated.

Uc

With transistors, the usual situation is for U_{CBO}(colletor base reverse bias) to be quoted, or U_{CEO} and U_{CEO} (collector emitter reverse bias). With FETs, U_{DS} (drain source voltage) is always quoted.

Ic

With transistors, I_C (collector current) is always quoted. If this is followed by (ss) in brackets, I_{CM} is quoted, i.e. the peak value of the collector current. With FETs, I_D (drain current) is always quoted.

Ptot

As a rule, the total leakage power Ptot is quoted, with RF types we always quote the RF output power P_Q, with corresponding frequency in brackets.

Amplification

The DC current gain B(h_{FE}) or the short-circuit current gain ß(h_{fe}) are always quoted as guaranteed values.

f_T

The transition frequency is always qouted in MHz.

Specifications of the transistors used in the projects

Type	Mat.	Pol.	Description	UC [Vmax]	IC [Amax]	Ptot [Wmax]	Current Gain	fT [MHz]
MJ3001	Si	npn	Darl+diode,pow	60	10.00	150.00($25°C)	>10	
MJE243	Si	npn	AF-s-pow	100	4.00	1.50($25°C)	40.120	>40.00
MJE244	Si	npn	AF-s-pow	100	4.00	1.50($25°C)	>25	>40.00
MJE253	Si	npn	AF-s-pow	100	4.00	1.50($25°C)	40-120	>40.00
MJE4350	Si	pnp	AF-end,s-pow	100	16.00	125.00($25°C)	15	>1.00
MJE5170	Si	pnp	uni-pow	120	6.00	2.00($25°C)	15-100	>1.00
MJE5180	Si	npn	uni-pow	120	6.00	2.00($25°C)	15-100	>1.00
MPF102	Si	n-ch	FET,VHF-in,sym,mix 25V, Idss>2mA,Up<V					
MPF106	Si	n-ch	FET,VHF 25V,Idss>4mA,Up<8V					
MPS-A29	Si	npn	Darl	100	0.50	1.50($25°C)	>10	>125.00
2N708	Si	npn	s	40/15	0.20	1.20(25°C)	>15	480.00
2N1711	Si	npn	uni	75	0.50	3.00(25°C)	75	>70.00
2N1889	Si	npn	AF-s	100/60	0.50	3.00(25°C)	40-120	>50.00
2N1890	Si	npn	AF-s	100/60	0.50	3.00(25°C)	100-300	>60.00
2N1990	Si	npn	nixie	100	1.00	2.00(25°C)	>25	
2N2102	Si	npn	AF-s	120/65	1.00	5.00(25°C)	40-120	>120.00
2N2222	Si	npn	ini	0		1.80(25°C)		
2N2368	Si	npn	fs	40/15	0.20	1.20(25°C)	20-60	>400.00
2N2369	Si	npn	fs	40/15	0.20	1.20(25°C)	40-120	>500.00
2N2905	Si	pnp	uni	60/40	0.60	3.00(25°C)	100-300	>200.00
2N2904	Si	pnp	uni	60/40	0.60	3.00(25°C)	40-120	>200.00
2N3019	Si	npn	uni	140/80	1.00	5.00(25°C)	100-300	>100.00
2N3020	Si	npn	uni	140/80	1.00	5.00(25°C)	40-120	>80.00
2N3055	Si	npn	AF-s-pow	100/60	15.00	115.00($25°C)	20-70	>2.50
2N3109	Si	npn	AF-s	80/40	1.00	5.00(25°C)	100-300	>70.00
2N3110	Si	npn	AF-s	80/40	1.00	5.00(25°C)	40-120	>60.00
2N3367	Si	n-ch	FET,uni,ln	40V,Idss>0.5mA,Up<2.5V				
2N3370	Si	n-ch	FET,uni,ln	40V,Idss>0.1mA,Up3.2V				
2N3454	Si	n-ch	FET,uni	50V,Idss>0.05mA,Up<2.3V				
2N3819	Si	n-ch	FET,VHF,uni,sym	25V,Idss>2mA,Up<8V				
2N3823	Si	n-ch	FET,VHF,ln	30V,Idss>4mA,Up<8V				
2N3903	Si	npn	uni	60/40	0.20	1.50(25°C)	50-150	>250.00
2N3904	Si	npn	uni	60/40	0.20	1.50(25°C)	100-300	>300.00
2N3905	Si	pnp	uni	40	0.20	1.50(25°C)	50-150	>200.0
2N3906	Si	pnp	uni	40	0.20	1.50(25°C)	100-300	>250.00
2N4118	Si	n-ch	FET,uni	40V,Idss>0.08mA,Up<3V				
2N5294	Si	npn	AF-s-pow	80/70	4.00	1.80($25°C)	30-120	>0.80
2N5397	Si	n-ch	FET,VHF/UHF	25V,Idss>10mA,Up<6V				
2N5398	Si	n-ch	FET,VHF/UHF	25V,Idss>5mA,Up<6V				
2N5486	Si	n-ch	FET,VHF/UHF	25V,Idss>8mA,Up<6V				
2N6038	Si	npn	Darl+diode,pow	60	4.00	1.50($25°C)	>10	>25.00
2N6039	Si	npn	Darl+diode,pow	80	4.00	1.50($25°C)	>10	>25.00
2N6283	Si	npn	Darl+diode,pow	80	20.00	160.00($25°C)	>10	>4.00
2N6284	Si	npn	Darl+diode,pow	100	20.00	160.0($25°C)	>10	>4.00
2N6412	Si	npn	AF-s-pow	60/40	4.00	15.00($25°C)	>5	>50.00
2N6414	Si	pnp	AF-s-pow	80/60	4.00	15.00($25°C)	>5	>50.00
2SA511	Si	pnp	AF/RF/s	90/80	1.50	8.00(25°)	30-150	60.00
2SA597	Si	pnp	RF-s	50/40	1.00	6.00($25°C)	10-250	400.00
2SA761	Si	pnp	uni	110	2.00	6.30($25°)	50-240	80.00
2SA970	Si	pnp	AF,ln	120	0.10	0.30(25°C)	200-700	100.0

Specifications of the transistors used in the projects

Type	Mat.	Pol.	Description	UC [Vmax]	IC [Amax]	Ptot [Wmax]	Current Gain	fT [MHz]
2SA1016	Si	pnp	uni,ln	120/100	0.05	0.40(25°)	160-960	110.00
2SA1123	Si	pnp	uni,ln	150	0.05	0.7(25°)	65-450	200.00
2SA1136	Si	pnp	AF-in,ln	120/100	0.10	0.30(25°C)	120-560	90.00
2SA1137	Si	pnp	AF-in,on	80	0.10	0.30(25°C)	120-560	90.00
2SA1141	Si	pnp	AF/Rf-pow	115	10.00	2.00($25°C)	100	80.00
2SA1285	Si	pnp	uni	120	0.20	0.90(25°C)	150-800	200.00
2SA1285A	Si	pnp	uni	150	0.10	0.90(25°C)	150-500	200.00
2SA1515	Si	pnp	uni	40/32	1.00	0.50(25°C)	82-390	150.00
2SA1705	Si	pnp	AF,s	60/50	1.00	0.90(25°C)	>30	150.00
2SA1706	Si	pnp	AF-s	60/50	2.00	1.00(25°C)	>40	150.00
2SB633	Si	pnp	AF-s-pow	100/85	6.00	40.00($25°C)	40-320	15.00
2SB764	Si	pnp	uni	60/50	1.00	0.90(25°C)	60-320	150.00
2SB822	Si	pnp	Af-dr/end	40/32	2.00	0.75(25°C)	82-390	100.00
2SB826	Si	pnp	s-pow	60/50	7.00	60.00($25°C)	>30	10.00
2SB867	Si	pnp	AF/s-pow,lo-sat	130/80	3.00	30.00($25°C)	60-260	30.00
2SB868	Si	pnp	AF/s-pow,lo-sat	130/80	4.00	35.00($25°C)	60-260	30.00
2SB869	Si	pnp	AF/s-pow,lo-sat	130/80	5.00	40.00($25°C)	60-260	30.00
2SB870	Si	pnp	AF/s-pow,lo-sat	120/80	7.00	40.00($25°C)	60-260	30.00
2SB874	Si	pnp	AF/s-pow, TV-VD	100/60	2.00	20.00($25°C)	>40	250.00
2SB909	Si	pnp	AF-dr/end	40/32	1.00	1.00(25°C)	82-390	150.00
2SB911	Si	pnp	AF-dr/end	40/32	2.00	1.00(25°C)	82-390	100.0
2SB920	Si	pnp		120/80				
2SB921	Si	pnp		120/80				
2SB1064	Si	pnp	AF-s-pow	60/50	3.00	1.50($25°)	60-320	70.00
2SB1114	Si	pnp	min,uni	20	2.00	2.00($25°C)	135-600	180.00
2SB1116	Si	pnp	uni	60/50	1.00	0.75(25°C)	135-600	120.00
2SB1142	Si	pnp	s-pow	60/50	2.50	10.00(25°C)	>35	140.00
2SB1143	Si	pnp	s-pow	60/50	4.00	10.00(25°C)	>40	150.00
2SB1144	Si	pnp	AF/s-pow,lo-sat	120/100	1.50	10.00(25°C)	>30	100.00
2SB1230	Si	pnp	AF/s-pow,lo-sat	110/100	15.00	100.00($25°C)	50-140	
2SB1231	Si	pnp	AF/s-pow,lo-sat	110/100	25.00	120.00($25°C)	50-140	
2SB1232	Si	pnp	AF/s-pow,lo-sat	110/100	40.00	150.00($25°C)	50-140	
2SC270	Si	npn	s-pow	270/75	5.00	50.00($25°C)	24-40	22.00
2SC460	Si	npn	AM-in/mix/osc	30	0.10	0.20(25°C)	35-200	230.00
2SC696	Si	npn	uni	100/60	3.00	0.75(25°C)	30-173	100.00
2SC763	Si	npn	VHF	25/12	0.02	0.10(25°C)	20-300	>400.00
2SC829	Si	npn	AM/FM-in/mix/osc	30/20	0.03	0.40(25°C)	40-500	230.00
2SC959	Si	npn	uni	120/80	0.70	0.70(25°C)	40-200	100.00
2SC1324	Si	npn	UHF-CATV	35/25	0.15	3.00(25°C)	10-35	
2SC1876	Si	npn	Darl	100/70	0.50	0.80(25°C)	>20	
2SC2124	Si	npn	TV-HD	220/800	2.00	5.00($90°C)	20	4.00
2SC2125	Si	npn	TV-HD	220/800	5.00	50.00($25°C)	8-25	5.00
2SC2270	Si	npn	lo-sat	50/20	5.00	1.00($25°C)	>70	100.00
2SC2334	Si	npn	s-pow,dc-dc conv.	150/100	7.00	40.00($25°C)	>20	
2SC2459	Si	npn	uni	120	0.10	0.20(25°C)	200-700	100.00
2SC2675	Si	npn	AF,ln	80	0.10	0.30(25°C)	180-820	120.00
2SC2724	Si	npn	FM-IF	30/25	0.03	0.20(25°C)	25-300	200.00
2SC3112	Si	npn	AF,ln	50	0.15	0.40(25°C)	600-3600	250.00
2SC3179	Si	npn	AF-pow	80/60	4.00	30.00($25°C)	100	15.00
2SC3245	Si	npn	uni	120	0.10	0.90(25°C)	150-800	200.00

Specifications of the transistors used in the projects

Type	Mat.	Pol.	Description	UC [Vmax]	IC [Amax]	Ptot [Wmax]	Current Gain	fT [MHz]
2SC3245A	Si	npn	uni	150	0.10	0.90(25°C)	400-800	200.00
2SC3248	Si	npn	uni	180	0.10	0.90(25°C)	150	130.00
2SC3358	Si	npn	UHF	20/12	0.10	0.25(25°C)	50-300	7000.00
2SC3420	Si	npn	lo-sat	50/20	5.00	10.00(25°C)	>70	100.00
2SC3622	Si	npn	AF-s,hi-beta	60/50	0.15	0.25(25°C)	1000-3200	250.00
2SC4308	Si	npn	VHF-A	30/20	0.30	0.60(25°C)	50-200	2500.00
2SD386	Si	npn	TV-VD	200/120	3.00	1.75($25°C)	40-320	8.00
2SD406	Si	npn	Darl	100	2.00	15.00(25°C)	>2000	
2SD613	Si	npn	AF-s-pow	100/85	6.00	40.00($25°C)	40-320	15.00
2SD614	Si	npn	Darl	100/80	3.00	0.80(25°C)	3000	15.00
2SD621	Si	npn	TV_HD	2500/900	3.00	50.00($25°C)	3-15	
2SD628	Si	npn	Darl+diode,pow	100	10.00	80.00($25°C)	>1000	
2SD629	Si	npn	Darl+diode,pow	100	10.00	100.00($25°C)	>1000	
2SD688	Si	npn	Darl,pow	100	1.50	0.80($25°C)	>10	
2SD712	Si	npn	AF-s-pow	100	4.00	30.00($25°C)	55-300	8.00
2SD726	Si	npn	AF-s-pow	100/80	4.00	40.00($25°C)	35-320	10.00
2SD729	Si	npn	Darl+diode,pow	100	20.00	125.00($25°C)	>1000	
2SD781	Si	npn	s-pow,TV-HD	150/60	2.00	1.00(25°C)	150	
2SD826	Si	npn		60/20	5.00	1.00($25°C)	120-560	120.00
2SD838	Si	npn	TV-HD,s-pow	2500/900	3.00	50.00($25°C)	3-15	
2SD892A	Si	npn	Darl	60/50	0.50	0.40(25°C)	>8000	150.00
2SD1049	Si	npn	AF-s-pow	120/80	25.00	80.00($25°C)	>20	
2SD1062	Si	npn	s-pow	60/50	12.00	40.00($25°C)	>30	10.00
2SD1153	Si	npn	Darl	80750	1.50	0.90(25°C)	>40	120.00
2SD1177	Si	npn	AF-pow,TV-HD	100/60	2.00	20.00($25°C)	>40	230.00
2SD1237	Si	npn	s-pow	90/80	7.00	1.75($25°C)	>30	20.00
2SD1238	Si	npn	s-pow	90/80	12.00	80.00($25°C)	>30	20.00
2SD1639	Si	npn	AF-s-pow	100/80	2.20	10.00($25°C)	40-200	
2SD1684	Si	npn	AF/s-pow,lo-sat	120/100	1.50	10.00(25°C)	>30	120.00
2SD1685	Si	npn	AF/s-pow,lo-sat	60/20	5.00	10.00(25°C)	>95	120.00
2SD1691	Si	npn	AF-s-pow	60	5.0	20.00(25°C)	100-400	
2SD1840	Si	npn	AF/s-pow,lo-sat	110/100	15.00	100.00($25°C)	50-140	
2SD1841	Si	npn	AF/s-pow,lo-sat	110/100	25.00	120.00($25°C)	50-140	
2SD1842	Si	npn	AF/s-pow,lo-sat	110/100	40.00	150.00($25°C)	50-140	
2SD2116	Si	npn	Darl	80/50	0.70	1.00(25°C)	>40	
2SD2117	Si	npn	Darl	80/50	1.50	1.00(25°C)	>30	
2SD2213	Si	npn	Darl,AF	150/80	1.50	0.90(25°C)	>10	
2SJ165	V-MOS	p-ch	FET-enh.,	50V,0.1A,0.25W				
2SK422	V-MOS	n-ch	FET-enh.	60v,0.7A,0.9W,17/12ns				
2SK423	V_MOS	n-ch	FET-enh.	100V,0.5A,0.9W,15/20ns				
3N140	MOS	n-ch	FET-depl.,dg,FM/VHF-in	20V,Idss>5mA				
3N225	MOS	n-ch	FET-depl.,dg, UHF	25V,Idss>1mA,Up<4V				
3SK35	MOS	n-ch	FET-depl.,dg,VHF	20V,Idss>3mA,Up<4V				
3SK37	MOS	n-ch	FET-depl.,dg,VHF	20V,Idss>4mA,Up<3V				
3SK45	MOS	n-ch	FET-depl.,dg,VHF	22V,Idss>4mA,Up<3V				
3SK61	MOS	n-ch	FET-depl.,dg,VHF	20V,Idss>4mA,Up<3V				
3SK72	MOS	n-ch	FET-depl.,dg,VHF	20V,Idss>2.5mA,Up<3V				
3SK77	MOS	n-ch	FET-depl.,dg,VHF	20V,Idss>3mA,Up<2.5V				
3SK85	MOS	n-ch	FET-depl.,dg,VHF	20V,Idss>4mA,Up<3V				

SEMICONDUCTOR DIODE SPECIFICATIONS

* RFR = Rectifier, Fast Recovery

Device	Type	Material	Peak Inverse Voltage,PIV (Volts)	Average Rectified Current Forward (Reverse) IO (A) (IR(A))	Peak Surge Current, IFSM 1 sec. @ 25°C (A)	Average Forward Voltage, VF (Volts)
1N34	Signal	Germanium	60	8.5 m (15.0µ)		1.0
1N34A	Signal	Germanium	60	5.0 m (30.0µ)		1.0
1N67A	Signal	Germanium	100	4.0 m (5.0µ)		1.0
1N191	Signal	Germanium	90	5.0 m	1.0	
1N270	Signal	Germanium	80	0.2 (100 µ)		1.0
1N914	Fast Switch	Silicon (Si)	75	75.0 m (25.0 n)	0.5	1.0
1N1184	RFR	Si	100	35 (10 m)		1.7
1N2071	RFR	Si	600	0.75 (10.0µ)		0.6
1N3666	Signal	Germanium	80	0.2 (25.0µ)		1.0
1N4001	RFR	Si	50	1.0 (0.03 m)		1.1
1N4002	RFR	Si	100	1.0 (0.03 m)		
1N4003	RFR	Si	200	1.0 (0.03 m)		1.1
1N4004	RFR	Si	400	1.0 [0.03 m)		1.1
1N4005	RFR	Si	600	1.0 (0.03 m)		1.1
1N4006	RFR	Si	800	1.0 (0.03 m)		1.1
1N4007	RFR	Si	1000	1.0 (0.03 m)		1.1
1N4148	Signal	Si	75	10.0 m (25.0 n)		1.0
1N4149	Signal	Si	75	10.0 m (25.0 n)		1.0
1N4152	Fast Switch	Si	40	20.0 m (0.05µ)		0.8
1N4445	Signal	Si	100	0.1 (50.0 n)		1.0
1N5400	RFR	Si	50	3.0	200	
1N5401	RFR	Si	100	3.0	200	
1N5402	RFR	Si	200	3.0	200	
1N5403	RFR	Si	300	3.0	200	
1N5404	RFR	Si	400	3.0	200	
1N5405	RFR	Si	500	3.0	200	
1N5406	RFR	Si	600	3.0	200	
1N5767	Signal	Si		0.1 (1.0µ)		1.0
ECG5863		RFR	Si	600 6	150	0.9

ZENER DIODES SPECIFICATIONS

Zener Voltage (Volts)	Power (Watts) 0.25	0.4	0.5	1.0	1.5	5.0	10.0	50.0
1.8	1N4614							
2.0	1N4615							
2.2	1N4616							
2.4	1N4617	1N4370,A	1N4370,A,1N5221,B					
			1N5985,B					
2.5			1N5222B					
2.6	1N702,A							
2.7	1N4618	1N4371,A	1N4371,A,1N5223,B					
			1N5839, 1N5986					
2.8			1N5224B					
3.0	1N4619	1N4372,A	1N4372,1N5225,B					
			1N5987					
3.3	1N4620	1N746,A	1N746A	1N3821	1N5913	1N5333,B		
		1N764 A	1N5226,B	1N4728,A				
		1N5518	1N5988					
3.6	1N4621	1N747,A	1N747A	1N3822	1N5914	1N5334,B		
		1N5519	1N5227,B,1N5989	1N4729,A				
3.9	1N4622	1N748,A	1N748A,1N5228,B	1N3823	1N5915	1N5335,B	1N3993A	1N4549,B
		1N5520	1N5844, 1N5990	1N4730,A				1N4557,B
4.1	1N704,A							
4.3	1N4623	1N749,A	1N749,A 1N5229,B	1N3824	1N5916	1N5336,B	1N3994,A	1N4550,B
		1N5521	1N5845,1N5991	1N4731 ,A				1N4558,B
4.7	1N4624	1N750,A	1N750A ,1N5230,B	1N3825	1N5917	1N5337,B	1N3995,A	1N4551,B
		1N5522	1N5846, 1N5992	1N4732,A				1N4559,B
5.1	1N4625	1N751 A	1N751A, 1N5231,B	1N3826	1N5918	1N5338,B	1N3996,A	1N4552,B
	1N4689	1N5523	1N5847,1N5993	1N4733		1N4560,B		
5.6	1N708A	1N752,A	1N752,A,1N5232,B	1N3827	1N5919	1N5339,B	1N3997,A	1N4553,B
	1N4626	1N5524	1N5848, 1N5994	1N4734,A		1N4561,B		
5.8	1N706A	1N762						
6.0				1N5233B			1N5340,B	
				1N5849				
6.2	1N709,1N4627	1N753,A	1N753,A	1N3828,A	1N5920	1N5341,B	1N3998,A	1N4554,B
	MZ605,MZ610	1N821,3,5,	1N5234,B, 1N5850	1N4735,A		1N4562,B		
	MZ620,MZ640	7,9; A	1N5995					
6.4	1N4565-84,A							
6.8	1N4099	1N754,A	1N754,A 1N757,B	1N3016,B	1N3785	1N5342,B	1N2970,B	1N2804B
		1N957,B	1N5235,B 1N5851	1N3829	1N5921		1N3999,A	1N3305B
		1N5526	1N5996	1N4736,A				1N4555,
								1N4563
7.5	1N4100	1N755,A	1N755A,1N958,B	1N3017,A,B	1N3786	1N5343,B	1N2971,B	1N2805,B
		1N958,B	1N5236,B, 1N5862	1N3830	1N5922	1N4000,A	1N3306,B	
		1N5527	1N5997	1N4737,A		1N4556,		
								1N4564
8.0	1N707A							
8.2	1N712A	1N756,A	1N756,A	1N3018,B	1N3787	1N5344,B	1N2972,B	1N2806,B
	1N4101	1N959,B	1N959,B,1N5237,B	1N4738,A	1N5923			1N3307,B
		1N5528	1N5853 ,1N5998					
8.4		1N3154-57,A	1N3154,A					
		1N3155-57						
8.5	1N4775-84,A		1N5238,B,1N5854					
8.7	1N4102					1N5345,B		
8.8		1N 764						
9.0		1N764A	1N935-9;A,B					

ZENER DIODES SPECIFICATIONS

Zener Voltage (Volts)	Power (Watts) 0.25	0.4	0.5	1.0	1.5	5.0	10.0	50.0
9.1	1N4103	1N757,A 1N960,B 1N5529	1N757,A, 1N960,B 1N5239,B, 1N5855 1N5999	1N3019,B 1N4739,A	1N3788 1N5924	1N5346,B	1N2973,B	1N2807,B 1N3308,B
10.0	1N4104	1N758,A 1N961,B 1N5530,B	1N758,A, 1N961,B 1N5240,B, 1N5856 1N6000	1N3020,B 1N4740	1N3789 1N5925	1N5347,B	1N2974,B	1N2808,B 1N3309,A,B
11.0	1N715,A 1N4105	1N962,B 1N5531	1N962,B,1N5241,B 1N5857, 1N6001	1N3021,B 1N4741,A	1N3790 1N5926	1N5348,B	1N2975,B	1N2809,B 1N3310,B
11.7	1N716,A 1N4106		1N941,A,B					
12.0		1N759,A 1N963,B 1N5532	1N759,A ,1N963,B 1N5242,B, 1N5858 1N6002	1N3022,B 1N4742,A	1N3791 1N5927	1N5349,B	1N2976,B	1N2810,B 1N3311,B
13.0	1N4107	1N964,B 1N5533	1N964,B,1N5243,B 1N5859,1N6003	1N3023,B 1N4743,A	1N3792 1N5928	1N5350,B	1N2977,B	1N2811,B 1N3312,B
14.0	1N4108	1N5534	1N5244B, 1N5860			1N5351,B	1N2978,B	1N2812,B 1N3313,B
15.0	1N4109	1N965,B 1N5535	1N965,B,1N5245,B 1N5861,1N6004	1N3024,B 1N4744A	1N3793 1N5929	1N5352,B	1N2979,A,B	1N2813,A,B 1N3314,B
16.0	1N4110	1N966,B 1N553,B	1N966,B,1N5246,B 1N5862, 1N6005	1N3025,B 1N4745,A	1N3794 1N5930	1N5353,B	1N2980,B	1N2814,B 1N3315,B
17.0	1N4111	1N5537	1N5247,B 1N5863			1N5354,B	1N2981B	1N2815,B 1N3316,B
18.0	1N4112	1N967,B 1N5538	1N967,B 1N5248,B 1N5864, 1N6006	1N3026,B 1N4746,A	1N3795 1N5931	1N5355,B	1N2982,B	1N2816,B 1N3917,B
19.0	1N4113	1N5539	1N5249,B 1N5865			1N5356,B	1N2983,B	1N2817,B 1N3318,B
20.0	1N4114	1N968,B 1N5540	1N968,B,1N5250,B 1N5866, 1N6007	1N3027,B 1N4747,A	1N3796 1N5932,A,B	1N5357,B	1N2984,B	1N2818,B 1N3319,B
22.0	1N4115	1N959,B 1N5541	1N969,B,1N5241,B 1N5867, 1N6008	1N3028,B 1N4748,A	1N3797 1N5933	1N5358,B	1N2985,B	1N2819,B 1N3320,A,B
24.0	1N4116	1N5542 1N9701B	1N970,B,1N5252,B 1N586,1N6009	1N3029,B 1N4749,A	1N3798 1N5934	1N5359,B	1N2986,B	1N2820,B 1N3321,B
25.0	1N4117	1N5543	1N5253,B 1N5869			1N5360,B	1N2987B	1N2821,B 1N3322,B
27.0	1N4118	1N971,B	1N971,1N5254,B 1N5870,1N6010	1N3030,B 1N4750,A	1N3799 1N5935	1N5361,B	1N2988,B	1N2822B 1N3323,B
28.0	1N4119	1N5544	1N5255,B,1N5871			1N5362,B		
30.0	1N4120	1N972,B 1N5546	1N972,B,1N5256,B 1N5872,1N6011	1N3031,B 1N4751,A	1N3800 1N5936	1N5363,B	1N2989,B	1N2823,B 1N3324,B
33.0	1N4121	1N973,B 1N5546	1N973,B,1N5257,B 1N5873,1N6012	1N3032,B 1N4752,A	1N3801 1N5937	1N5364,B	1N2990,A,B	1N2824,B 1N3325,B
36.0	1N4122	1N974,B	1N974,B,1N5258,B 1N5874,1N6013	1N3033,B 1N4753,A	1N3802 1N5938	1N5365,B	1N2991,B	1N2825,B 1N3326,B
39.0	1N4123	1N975,B	1N975,B, 1N5259,B 1N5875 ,1N6014	1N3034,B 1N4754,A	1N3803 1N5939	1N5366,B	1N2992,B	1N2826,B 1N3327,B
43.0	1N4124	1N976,B	1N976,B,1N5260,B 1N5876,1N6015	1N3035,B 1N4755,A	1N3804 1N5940	1N5367,B	1N2993,A,B	1N2827,B 1N3328,B
45.0			1N2994B	1N2828B 1N3329B				

POWER FETs

Device No.	Type	Max. Diss. (W)	Max. V_{DS} (Volts)	Max I_D (A)*	Gfs mmhos (typ.)	Input C_{iss} (pF)	Output C_{oss} (pF)	Approx. Upper Freq. (MHz)	Case	Package/ Type Mnfr.	General applications
DV1202S	N-Chan.	10	50	0.5	100k	14	20	500	.380 SOE	1/S	RF power amp., oscillator
DV1202W	N-Chan.	10	50	0.5	100k	14	20	500	C-220	5/S	RF power amp., oscillator
DV1205S	N-Chan.	20	50	1	200k	26	38	500	.380 SOE	1/S	RF power amp., oscillator
DV1205W	N-Chan.	20	50	1	200k	26	98	500	C-220	5/S	RF power amp., oscillator
2SK133	N-Chan.	100	120	7	1M	600	350	1	TO-3	6/H	AF pwr. amp., switch (complem to 25J48)
2SK134	N-Chan.	100	140	7	1M	600	350	1	TO-3	6/H	AF pwr. amp., switch (complem to 25J49)
2SK135	N-Chan.	100	160	7	1M	600	350	1	TO-3	6/H	AF pwr. amp., switch (complem to 25J50)
2SJ48	P-Chan.	100	120	7	1M	900	400	1	TO-3	6/H	AF pwr. amp., switch (complem to 2SK133)
2SJ49	P-Chan.	100	140	7	1M	900	400	1	TO-3	6/H	AF pwr. amp., switch (complem to 2SK134)
2SJ50	P-chan.	100	160	7	1M	900	400	1	TO-3	6/H	AF pwr. amp., switch (complem to 2SK135)
VMP4	N-Chan.	25	60	2	170K	32	4.8	200	.380 SOE	1/S	VHF pwr. amp., rcvr front end (rf amp., mixer).
VN10KM	N-Chan.	1	60	0.5	100K	48	16	-	TO-92	2/S	High-speed line driver, relay driver, LED stroke driver
VN64GA	N-Chan.	80	60	12.5	150K	700	325	30	TO-3	3/S	Linear amp., power-supply switch, motor control
VN66AF	N-Chan.	15	60	2	150K	50	50	-	TO-202	4/S	High-speed switch, HF linear amp., audio amp. line driver.
VN66AK	N-Chan.	8.3	60	2	250K	93	6	100	TO-39	7/S	RF pwr.amp.,high-current analog switching
VN67AJ	N-Chan.	25	60	2	250K	33	7	100	TO-3	3/S	RF pwr.amp.,high-current switching
VN89AA	N-Chan.	25	80	2	250K	50	10	100	TO-3	3/S	High-speed switching,HF linear amps., line drivers.
IRF100	N-Chan.	125	80	16	300K	900	25	-	TO-3	3/S	High-speed switching,audio inverters.
IRF101	N-Chan.	125	60	16	300K	900	25	-	TO-3	3/S	Same as IRF100

Legend: * 25°C (case) S = M/A-COM H = Hitachi IR = International Rectifier. Mnfr = Manufacturer

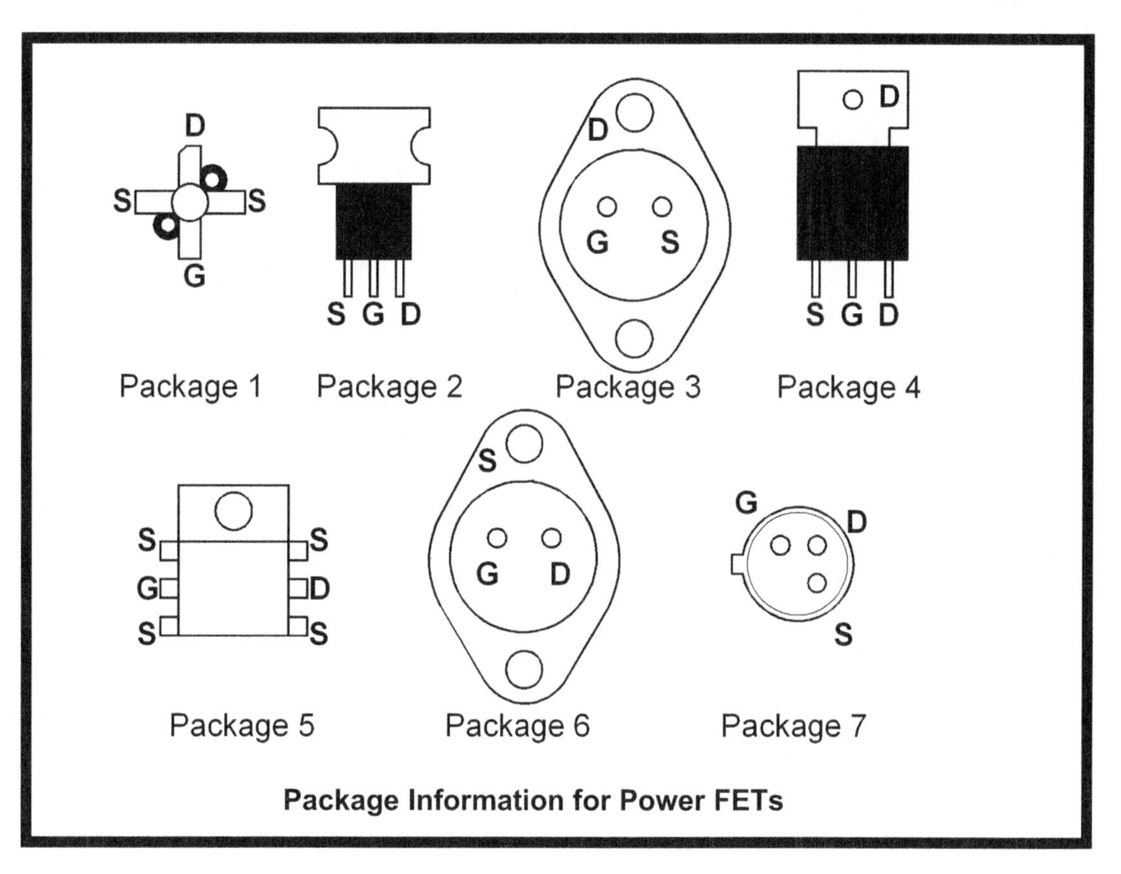

Package Information for Power FETs

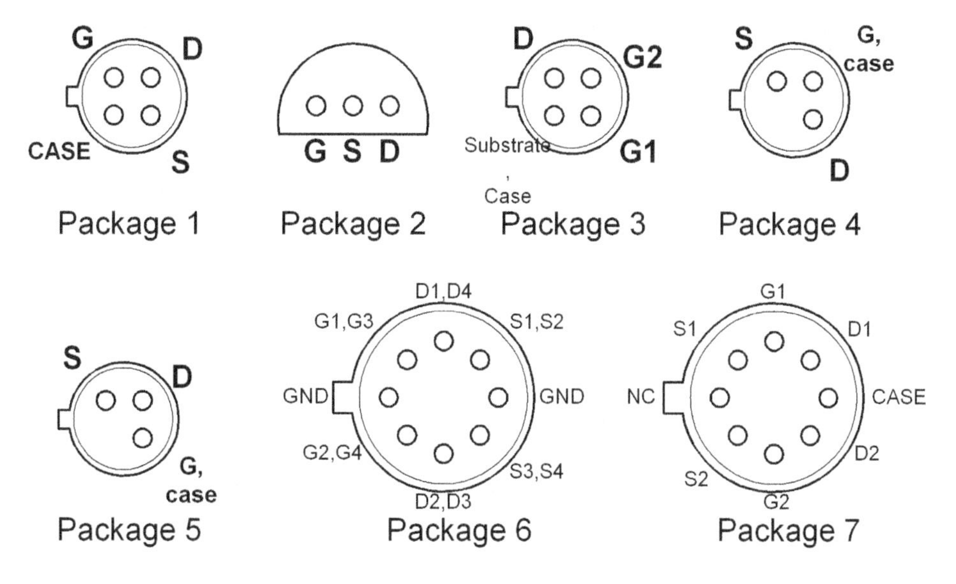

Package Information for Small Signal FETs

SMALL-SIGNAL FETs

Device No.	Type	Max. Diss. (mW)	Max. V_{DS} (Volts)	Max I_D	Min G_{fs} (mA)*	Input C (mS)	$V_{GS(off)}$ (pF)	Upper Freq. (volts)	Noise Figure (MHz)	Case Type (typ)	/Mnfr.	General applications
2N4416	N-JFET	300	30	-15	4.5K	4	-6	450	400 MHz 4 dB	TO-72	1/S,M	VHF/UHF/RF amp.mix., osc.
2N5484	N-JFET	310	25	30	2.5K	5	-3	200	200 MHz 4 dB	TO-92	2/M	VHF/UHFamp,mix., osc.
2N5485	N-JFET	310	25	30	3.5K	5	-4	400	400 MHz 4 dB	TO-92	2/S	VHF/UHF/RF amp.mix., osc.
3N200	N-Dual-Gate MOSFET	330	20	50	10K	4-8.5	-6	500	400 MHz 4.5 dB	TO-72	3/R	VHF/UHF/RF amp.mix., osc.
3N202	N-Dual-Gate MOSFET	360	25	50	8K	6	-5	200	200 MHz 4.5 dB	TO-72	3/S	VHF amp., mixer
MPF102	N-JFET	310	25	20	2K	4.5	-8	200		TO-92	2/N,M	HF/VHF amp.,mix., osc.,
MPF106/ 2N5484	N-JFET	310	25	30	2.5K	5	-6	400	200 MHz 4 dB	TO-92	2/N,M	HF/VHF/UHF amp.,mix.,osc.
40673	N-Dual-Gate MOSFET	330	20	50	12K	6	-4	400	200 MHz 6 dB	TO-72	3/R	HF/VHF/UHF amp. mix., osc.
U300	P-JFET	300	-40	20	8K	-50	+10	-	400 MHz	TO-18	4/S	General Purpose amp.
U304	P-JFET	350	-30	-50		27	+10	-	-	TO-18	4/S	analog switch, chopper
U310	N-JFET	500	30	60	10K	2.5	-6	450	450 MHz 3.2 dB	TO-52	5/S	common-gate VHF/UHF amp.,osc., mixer
	300	30										
U350	N-JFET Quad	1W	25	60	9K	5	-6	100	100 MHz 7 dB	TO-99	6/S	matched JFET doubly bal. mixer
U431	N-JFET Dual	300	25	30	10K	5	-6	100	-	TO-99	7/S	matched JFET cascade amp., balanced mixer

* 25°C   S = Siliconix Inc.   R = RCA   N = National Semiconductor   M = Motorola

Three-Terminal Voltage Regulators

* Listed numerically by device

Device	Description	Voltage	Current (Amps)	Package
317	Adj. Pos	+1.2 to +37	0.5	TO-205
317	Adj. Pos	+1.2 to +37	1.5	TO-204,TO-220
317L	Low Current Adj. Pos	+1.2 to +37	0.1	TO-205,TO-92
317M	Med Current Adj. Pos	+1.2 to +37	0.5	TO-220
350	High Current Adj. Pos	+1.2 to +33	3.0	TO-204,TO-220
337	Adj. Neg	-1.2 to -37	0.5	TO-205
337	Adj. Neg	-1.2 to -37	1.5	TO-204,TO-220
337M	Med Current Adj. Neg	-1.2 to -37	0.5	TO-220
309		+5	0.2	TO-205
309		+5	1.0	TO-204
323		+5	9.0	TO-204,TO-220
140-XX	Fixed Pos	**Note #**	1.0	TO-204,TO-220
340-XX			1.0	TO-204,TO-220
78XX			1.0	TO-204,TO-220
78LXX			0.1	TO-205,TO-92
78MXX			0.5	TO-220
78TXX			3.0	TO-204
79XX	Fixed Neg	**Note #**	1 .0	TO-204,TO-220
79LXX			0.1	TO-205,TO-92
79MXX			0.5	TO-220

Legend:

Adj.	= Adjustable
Med	= Medium
Neg	= Negative
Pos	= Positive

Note # - XX indicates the regulated voltage; which may be anywhere from 1.2 volts to 35 volts. For example a 7808 is a positive 8-volt regulator, and a 7912 is a negative 12-volt regulator.

The regulator package may be denoted by an additional suffix, according to the following:

Package	**Suffix**
TO-204 (TO-3)	K
TO-220	T
TO-205 (TO-39)	H,G
TO-92	P,Z

Example:

A 7815K is a positive 15-volt regulator in a TO-204 package. An LM340T-8 is a positive 8-volt regulator in a TO-220 package. In addition, different manufacturers use different prefixes. An LM7812 is equivalent to a μA 7812 or MC7812.

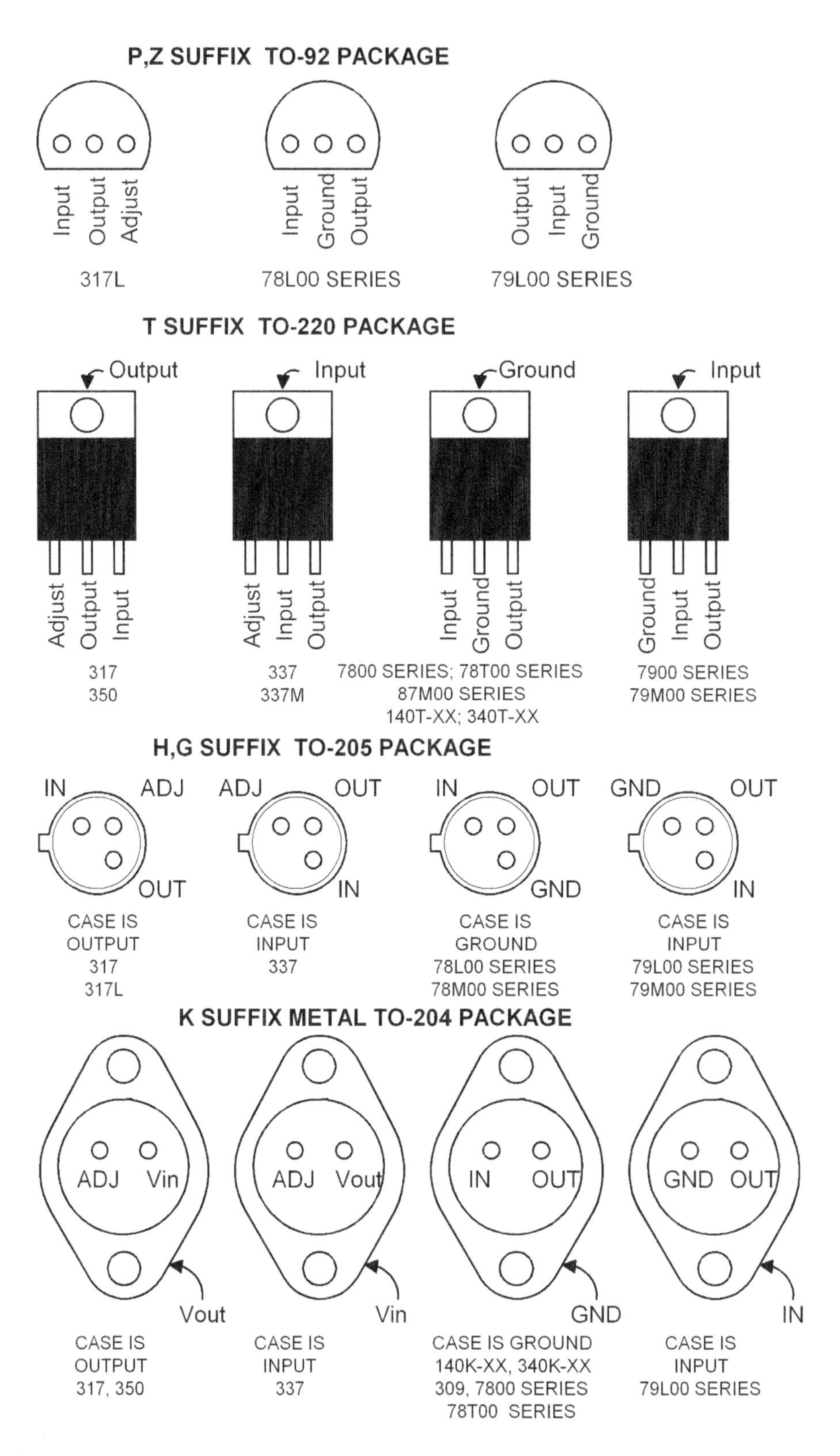
P,Z SUFFIX TO-92 PACKAGE
Input
Output
Adjust
317L
Input
Ground
Output
78L00 SERIES
Output
Input
Ground
79L00 SERIES
T SUFFIX TO-220 PACKAGE
Output
Adjust
Output
Input
317
350
Input
Adjust
Input
Output
337
337M
Ground
Input
Ground
Output
7800 SERIES; 78T00 SERIES
87M00 SERIES
140T-XX; 340T-XX
Input
Ground
Input
Output
7900 SERIES
79M00 SERIES
H,G SUFFIX TO-205 PACKAGE
IN
ADJ
OUT
CASE IS
OUTPUT
317
317L
ADJ
OUT
IN
CASE IS
INPUT
337
IN
OUT
GND
CASE IS
GROUND
78L00 SERIES
78M00 SERIES
GND
OUT
IN
CASE IS
INPUT
79L00 SERIES
79M00 SERIES
K SUFFIX METAL TO-204 PACKAGE
ADJ
Vin
Vout
CASE IS
OUTPUT
317, 350
ADJ
Vout
Vin
CASE IS
INPUT
337
IN
OUT
GND
CASE IS GROUND
140K-XX, 340K-XX
309, 7800 SERIES
78T00 SERIES
GND
OUT
IN
CASE IS
INPUT
79L00 SERIES

PRINTED CIRCUIT BOARD LAYOUTS

All printed circuit board layouts in this collection are once again printed in the following pages. You can either cut out or photocopy these pages to make a separate file for quick reference.

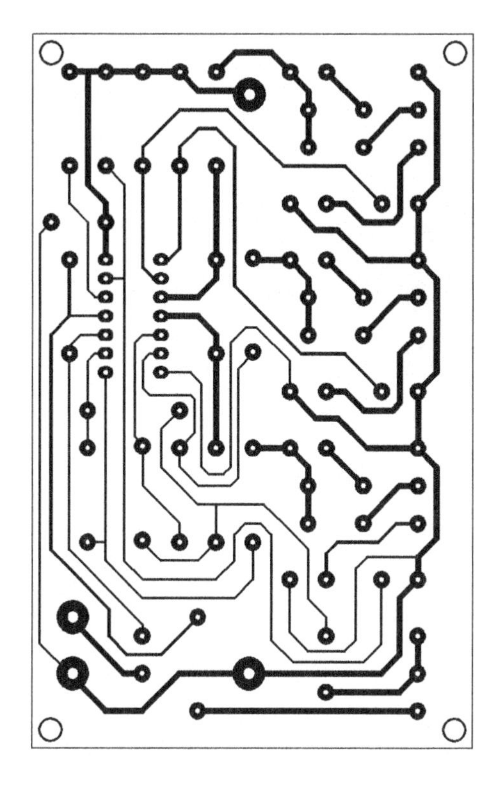

page 11 3-Channel Audio Mixer

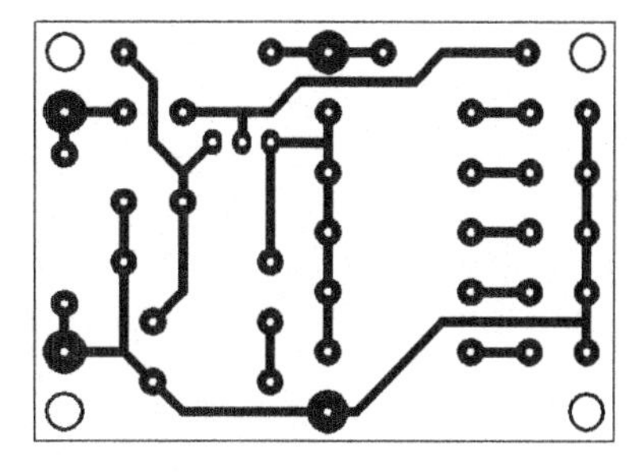

page 20 Audio Mixer

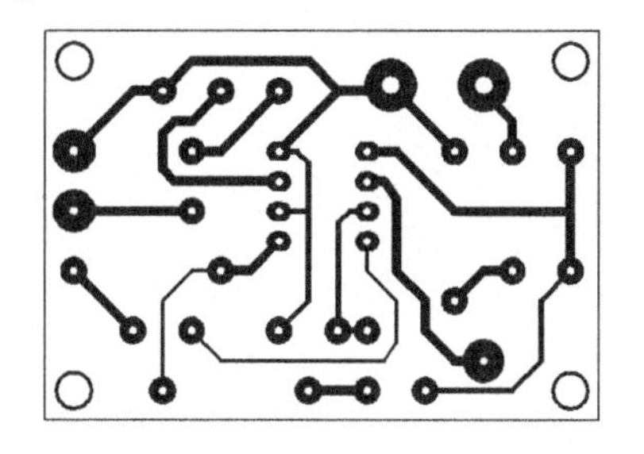

page 16 LM386 Audio Amp

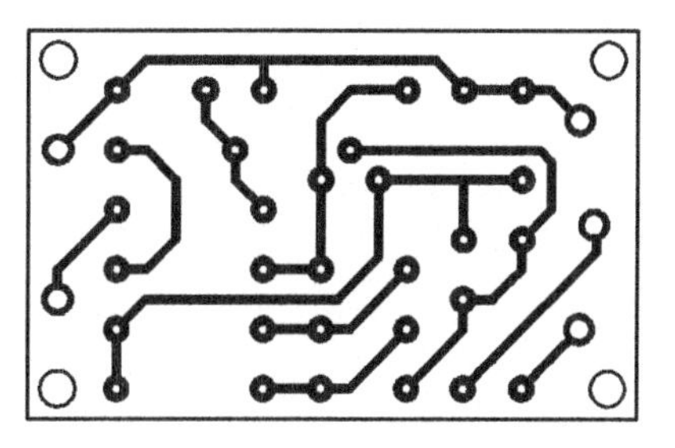

page 25 Audio Frequency Generator

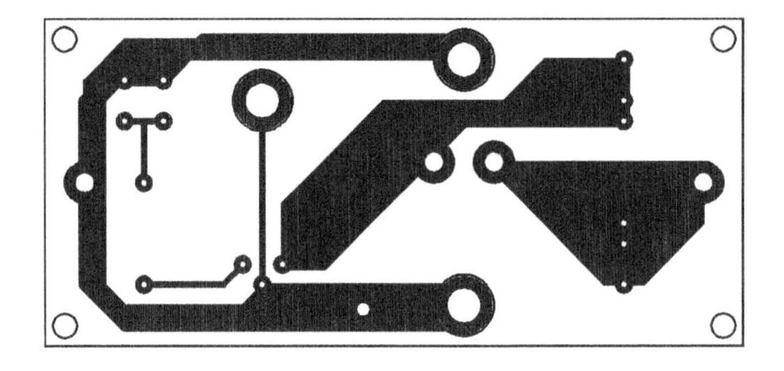

page 26 Electronic Fuse

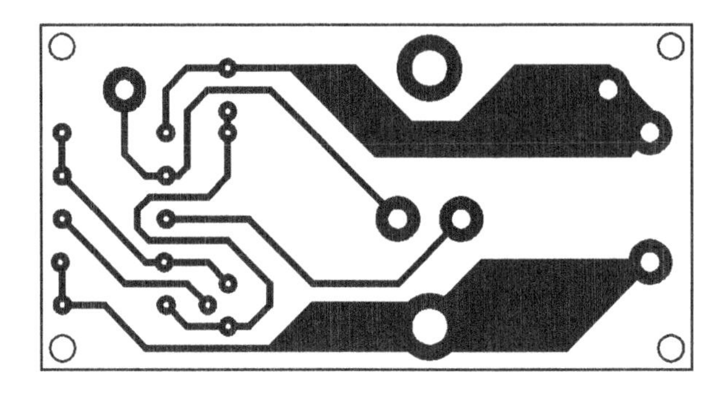

page 29 Blinker Circuit 1

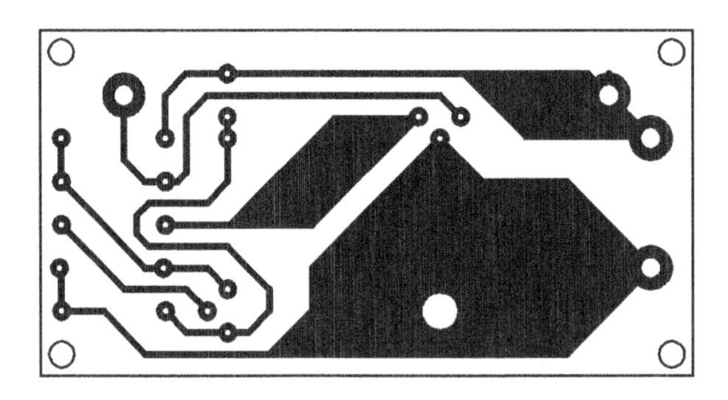

page 29 Blinker Circuit 2

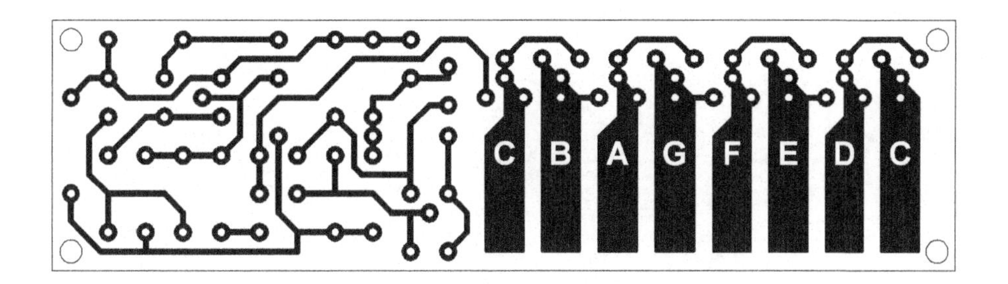

page 33 Electronic Organ

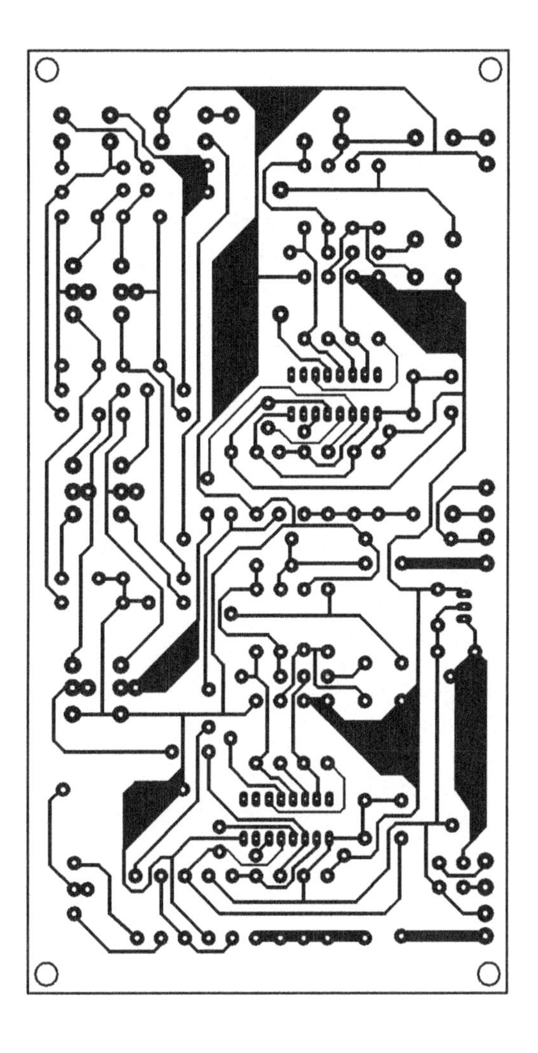

page 35 HIFI Stereo Preamp

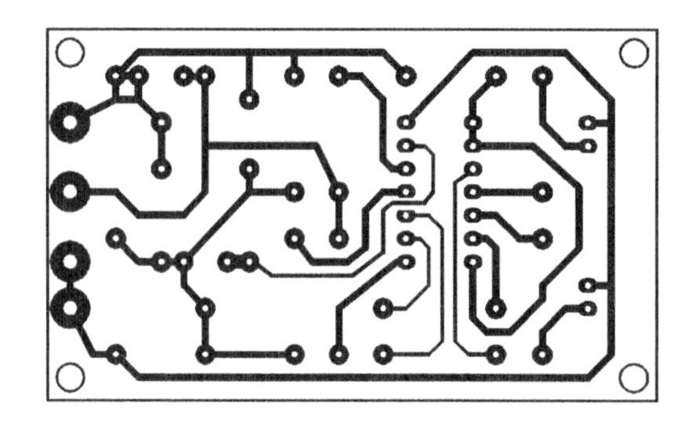

page 40 Temperature Monitor

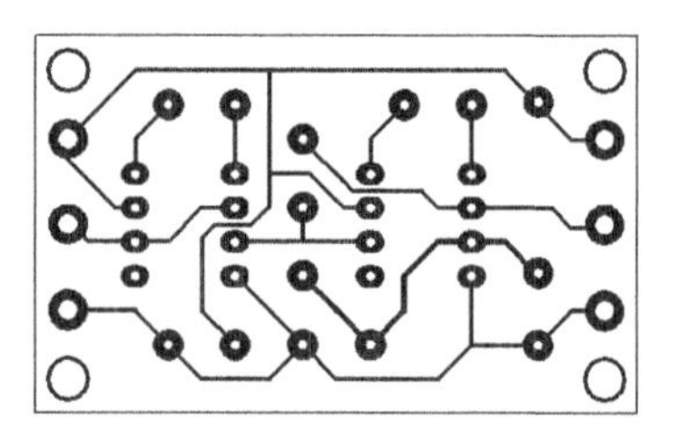

page 43 Touch Volume Control

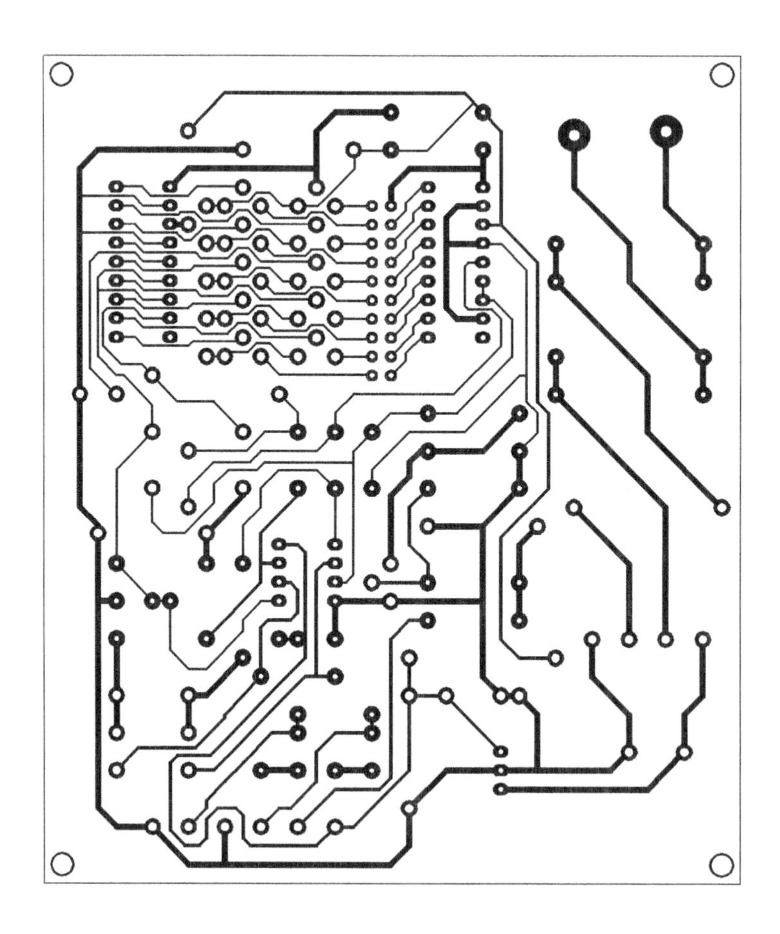

page 49 XY VU Display

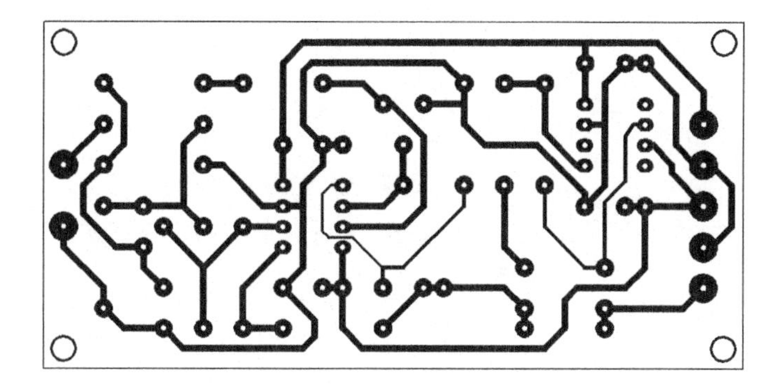

page 60 Automatic Volume Control

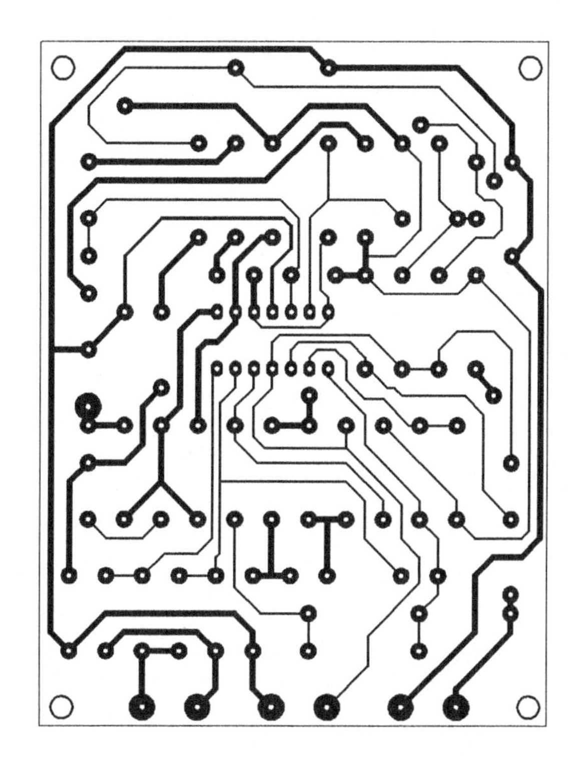

page 62 Audio Tester

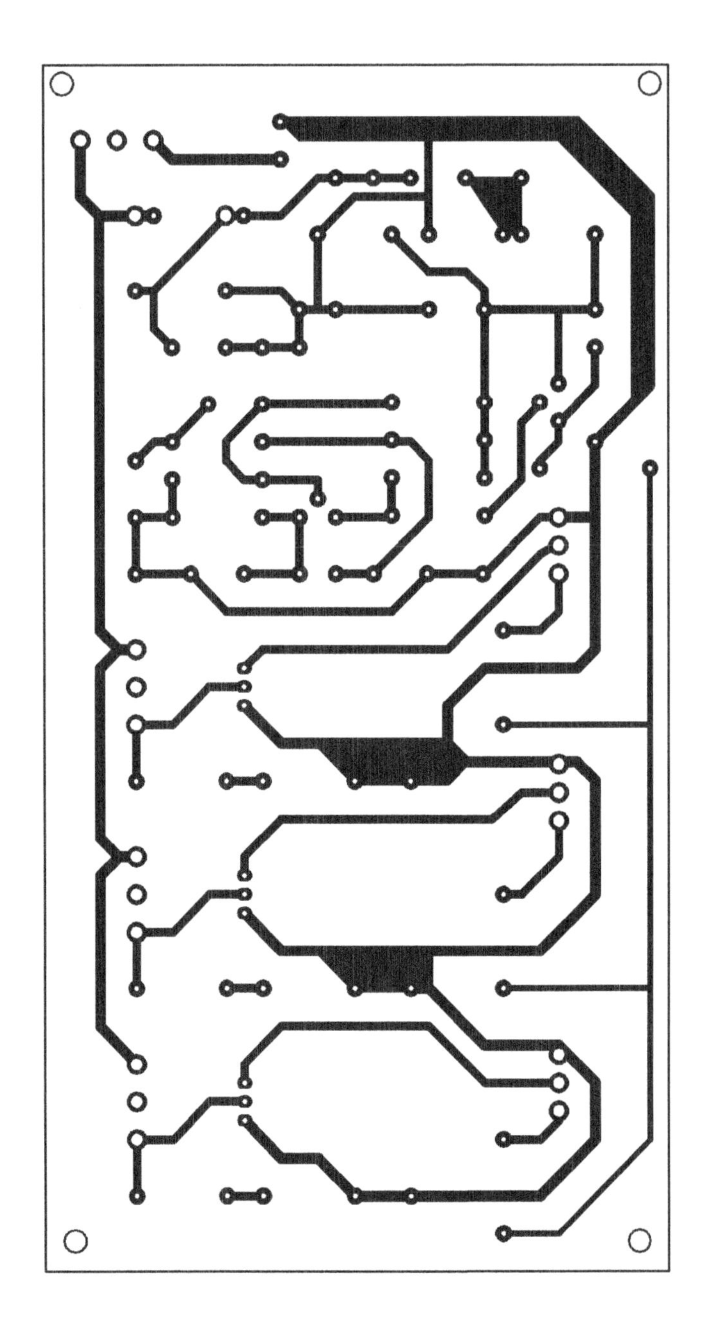

page 66 Sound Controlled Lights

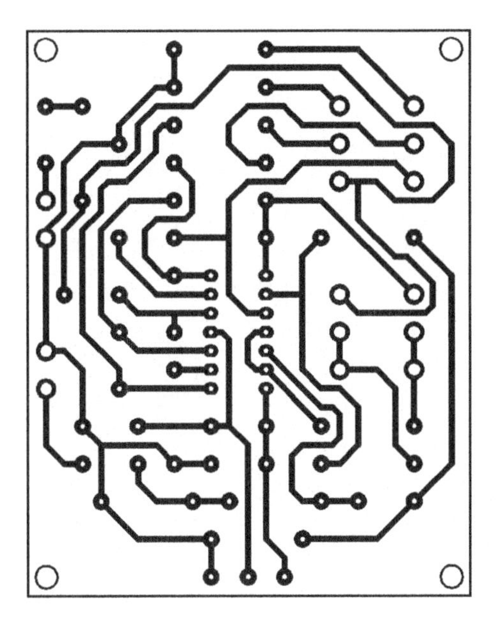

page 70 DX Audio Filter

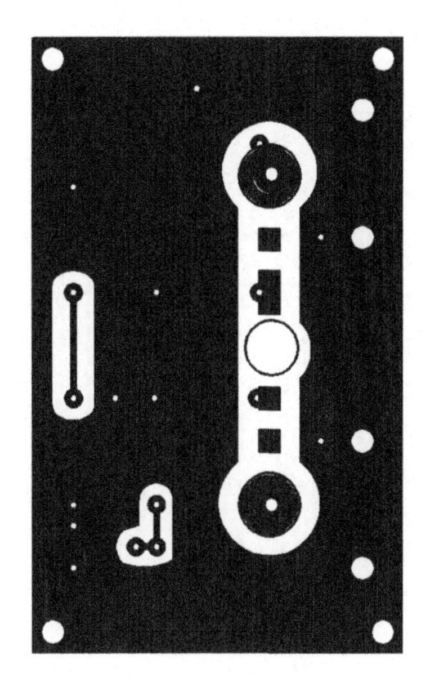

page 74 UHF Antenna Preamp

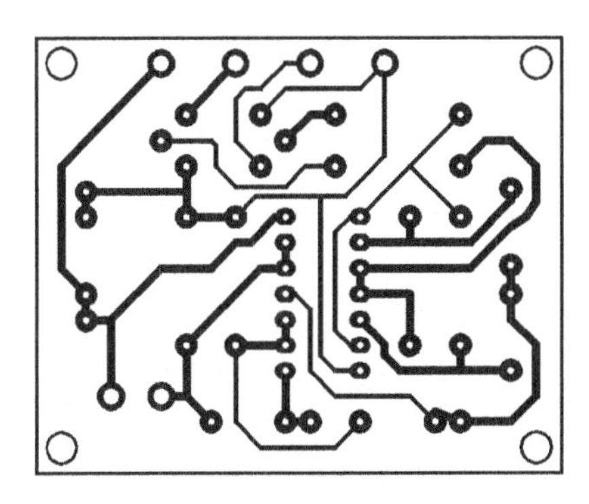

page 80 Electronic Head or Tail

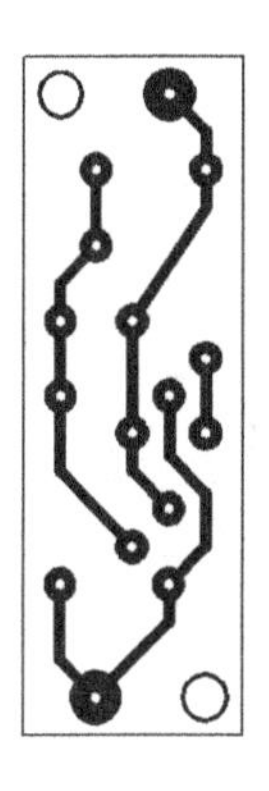

page 85
Adjustable Zener Circuit

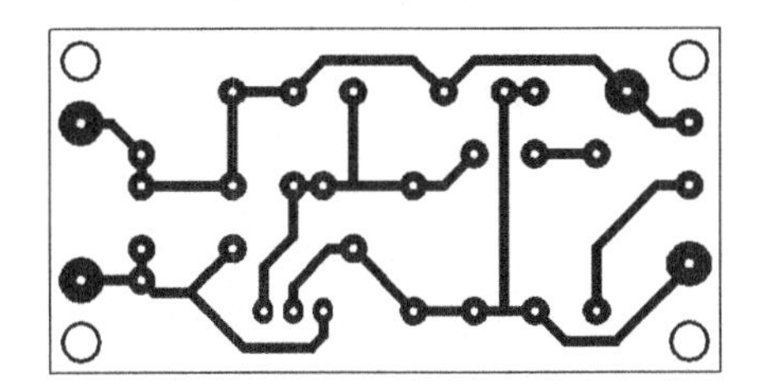

page 91 3-Ampere Power Supply

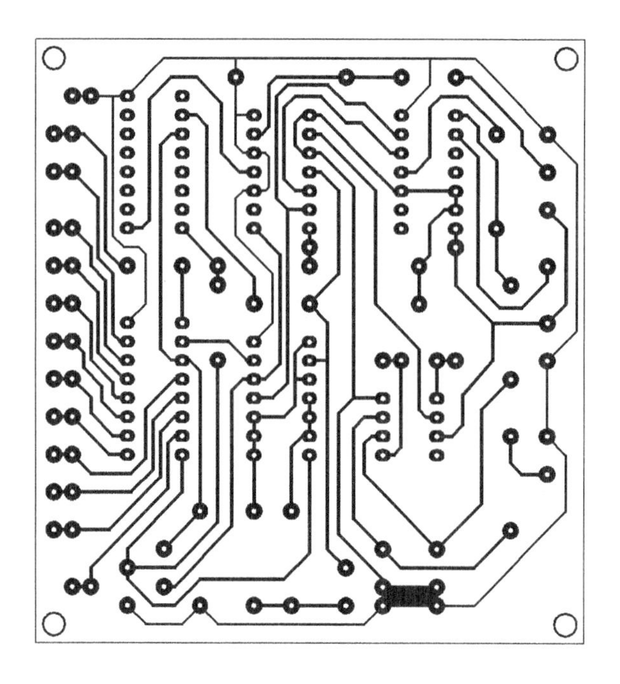

page 95 Analog-Digital Converter

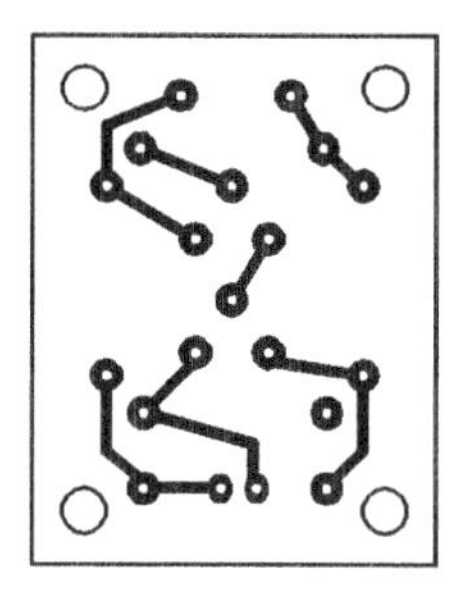

page 99
Fuse and Lamp Tester

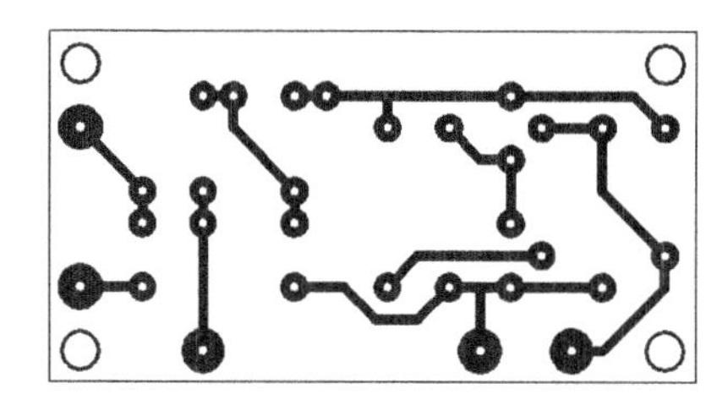

page 100 Low Current Relay

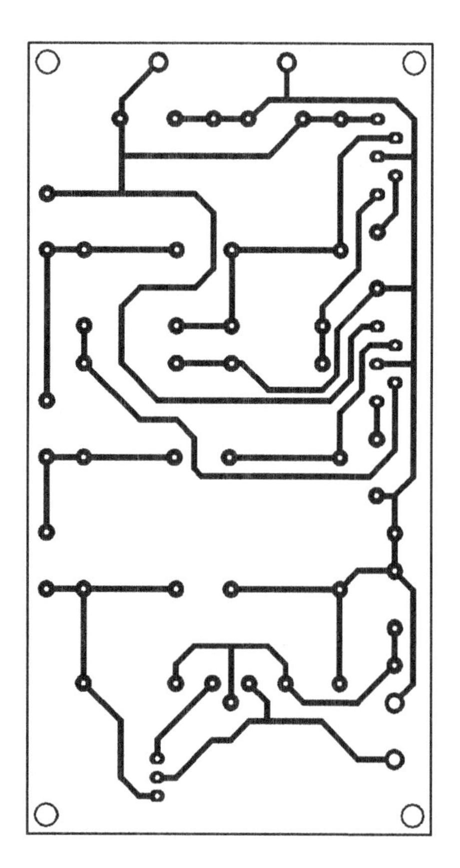

page 106 6V-12V Converter

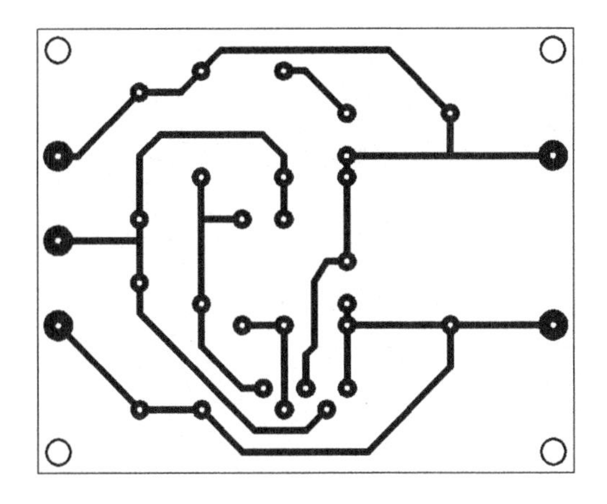

page 117 Symmetrical Power Supply

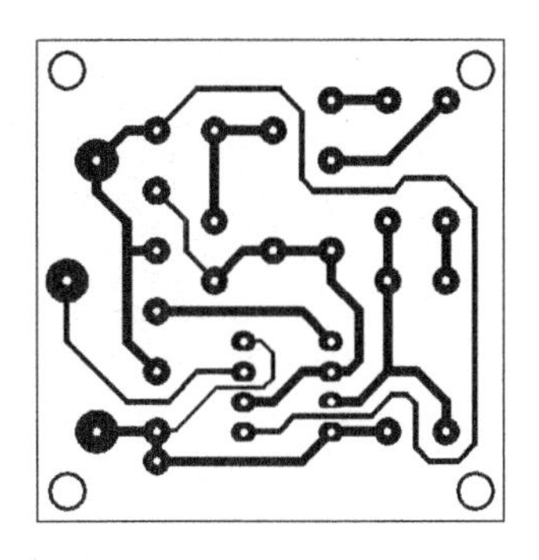

page 121 Pulse Generator

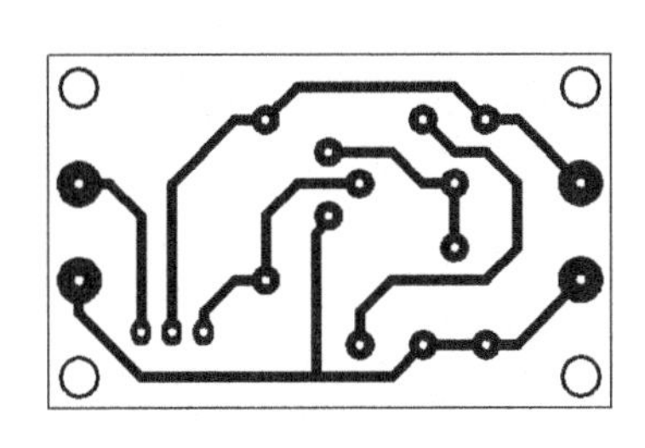

page 124 LED Dimmer

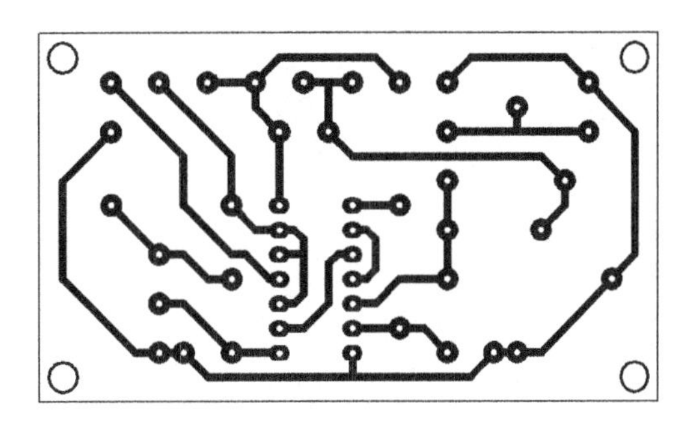

page 132 Refrigerator Alarm

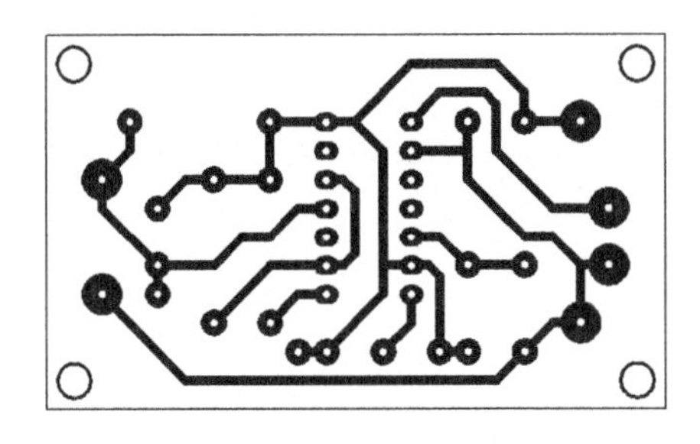

page 136 Water Level Monitor

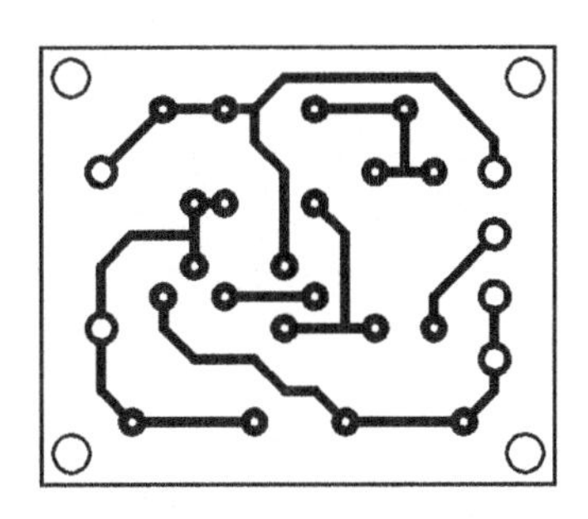

page 138 Siren

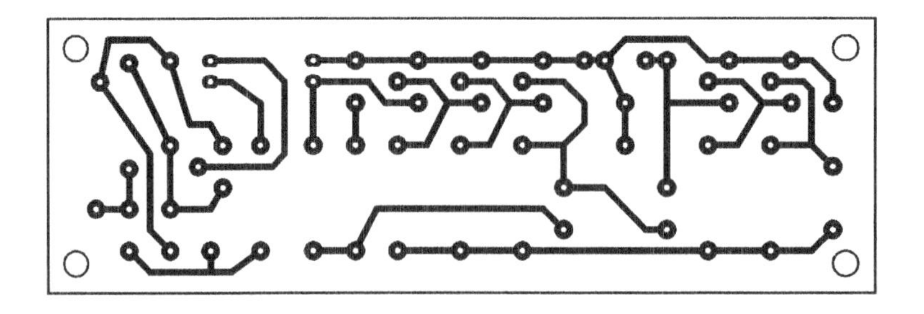

page 139 LED Optocoupler

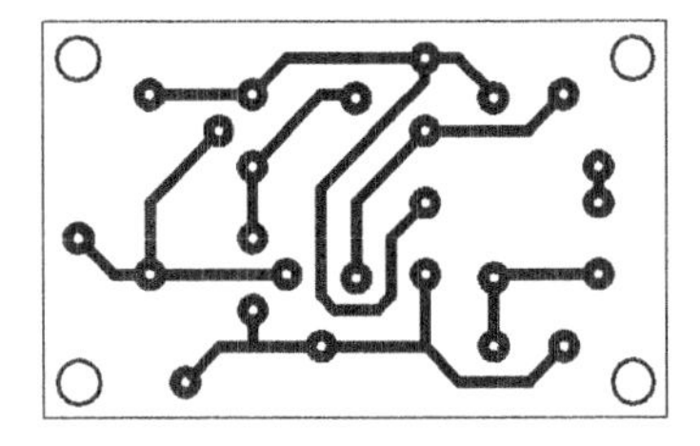

page 149 Pulse Duty Cycle Meter

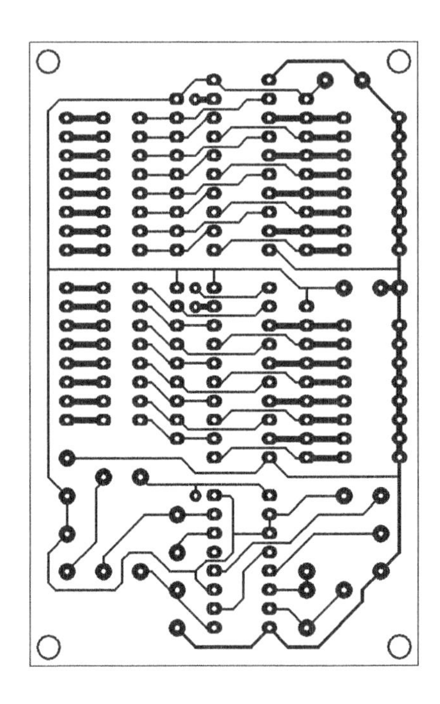

page 152 Digitally Controlled Trigger

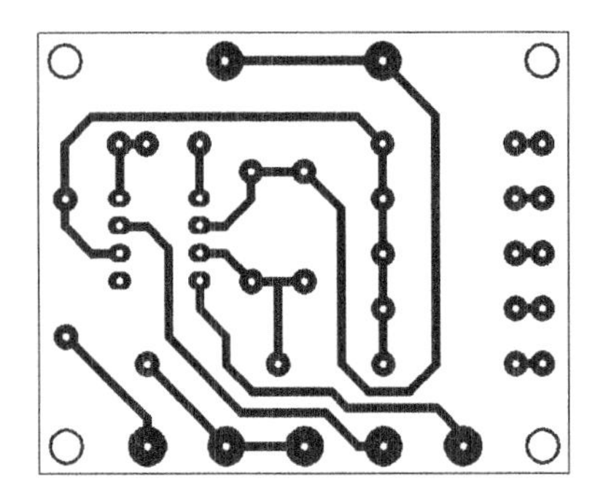

page 156 Microampere Meter

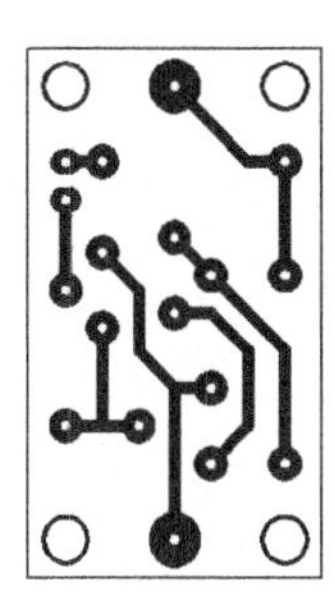

page 154 Temperature Monitor

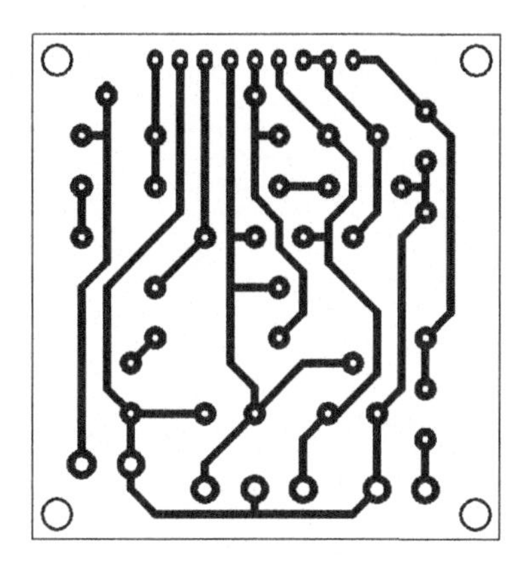

page 44
1-Chip 40 Watt Amplifier

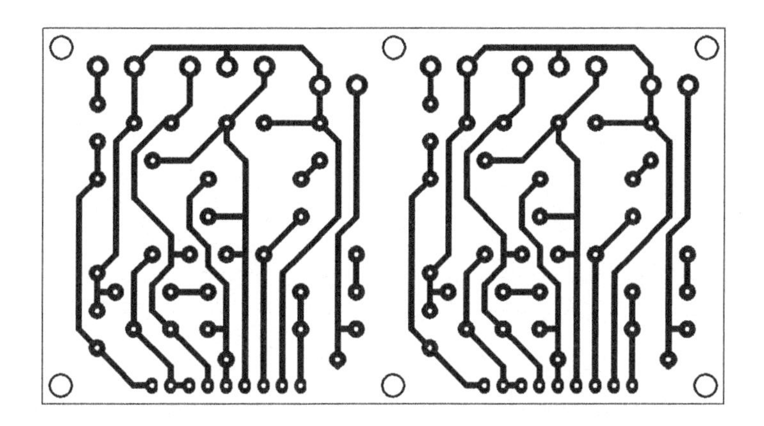

page 44 1-Chip 40 Watt Amplifier (stereo version)

Index

Index

Index

Index

Notes

Notes

Notes

Notes

Notes

Notes

Notes

Notes

ISBN 9781419690051
90000
9 781419 690051

Printed in Great Britain
by Amazon